# FLOW CYTOMETRY AND CELL SORTING

# FLOW CYTOMETRY AND CELL SORTING

*Edited by:*

**Prerna Pandey and Shiv Sanjeevi**

Delve PUBLISHING

www.delvepublishing.com

**Flow Cytometry and Cell Sorting**

*Prerna Pandey and Shiv Sanjeevi*

**Delve Publishing**
2010 Winston Park Drive,
2nd Floor
Oakville, ON L6H 5R7
Canada
www.delvepublishing.com
Tel: 001-289-291-7705
001-905-616-2116
Fax: 001-289-291-7601
Email: orders@arclereducation.com



ISBN: 978-1-77361-388-8 (Hardcover)

# DECLARATION

# ABOUT THE EDITORS

**Dr Prerna Pandey**, a biotechnologist with several years of wet lab research experience, has worked as a senior research fellow at International Center for Genetic Engineering and Biotechnology, New Delhi. Her field of research in PhD included isolation and molecular characterization of Geminiviruses, genome sequencing, gene annotation, as well as gene silencing using the RNA interference technology. She has published papers in peer-reviewed journals in the field and has a submitted a number of annotated Geminiviral genome sequences in the GenBank, including two novel ones. Prerna has also completed her editing and proof-reading courses from SfEP, UK and now works as a freelance scientific editor, writer and a subject matter expert. She has edited and co-authored several books on a range of topics in Life-Sciences. When Prerna is not busy with her assignments, she loves visiting historical places.

**Shiv Sanjeevi Sripathi** secured second rank in Mumbai University and first rank in his college at his undergrad level and in his college during his Masters and has worked on a stem cell project at the Specialized Centre for Cell Based Therapy (SCCT), KEM Hospital at Mumbai, following which he was a Junior Research Fellow at Microbiology & Cell Biology Department, Indian Institute of Science, Bangalore from August 2009 to November 2011. The field of work involved cloning of cell wall genes and transcription factors in E. coli & M. smegmatis. This was followed by a stint as a Research and Development Executive at Transasia Biomedicals, Mumbai where he worked on patent writing and Immunoassays from 2011 till 2013. In 2013, he entered the teaching field where he has been tutoring biology and chemistry to the students of 11th and 12th grades along with English grammar and communication skills to people of all age groups. Shiv also loves helping people write scientific manuscripts and has co-edited several books on life sciences and aspects of biotechnology with Apple Academic Press.

# TABLE OF CONTENTS

# LIST OF CONTRIBUTORS

**Shiv Sanjeevi Sripathi**

**Dr Prerna Pandey**

**Zhi** Han
College of Software, Nankai University, Tianjin, China
Department of Biomedical Informatics, The Ohio State University, Columbus, OH, USA

**Travis Johnson**
Department of Biomedical Informatics, The Ohio State University, Columbus, OH, USA

**Jie** Zhang
Department of Biomedical Informatics, The Ohio State University, Columbus, OH, USA
The CCC Biomedical Informatics Shared Resource, The Ohio State University, Columbus, OH, USA

**Xuan** Zhang
College of Software, Nankai University, Tianjin, China

**Kun Huang**
Department of Biomedical Informatics, The Ohio State University, Columbus, OH, USA

**Mary Beth Hanley**
Research and Development, BD Biosciences, San Jose, California, United States of America

**Woodrow Lomas**
Research and Development, BD Biosciences, San Jose, California, United States of America

**Dev Mittar**
Research and Development, BD Biosciences, San Jose, California, United States of America

**Vernon Maino**
Research and Development, BD Biosciences, San Jose, California, United States of America

**Emily Park**
Research and Development, BD Biosciences, San Jose, California, United States of America

**Peter C. Dumoulin**
W. Harry Feinstone Department of Molecular Microbiology and Immunology, Bloomberg School of Public Health, Johns Hopkins University, Baltimore, MD, 21205, United States of America

**Stefanie A. Trop**
W. Harry Feinstone Department of Molecular Microbiology and Immunology, Bloomberg School of Public Health, Johns Hopkins University, Baltimore, MD, 21205, United States of America

**Jinxia Ma**
W. Harry Feinstone Department of Molecular Microbiology and Immunology, Bloomberg School of Public Health, Johns Hopkins University, Baltimore, MD, 21205, United States of America

**Hao Zhang**
W. Harry Feinstone Department of Molecular Microbiology and Immunology, Bloomberg School of Public Health, Johns Hopkins University, Baltimore, MD, 21205, United States of America

**Matthew A. Sherman**
Triangle Research Labs, 6 Davis Drive, Durham, NC, 27709, United States of America

**Jelena Levitskaya**
W. Harry Feinstone Department of Molecular Microbiology and Immunology, Bloomberg School of Public Health, Johns Hopkins University, Baltimore, MD, 21205, United States of America

**Verónica Ambriz-Aviña**
Departamento de Biología, División de Ciencias Naturales y Exactas, Universidad de Guanajuato, Noria Alta S/N, 36050 Guanajuato, GTO, Mexico

**Jorge A. Contreras-Garduño**
Departamento de Biología, División de Ciencias Naturales y Exactas, Universidad de Guanajuato, Noria Alta S/N, 36050 Guanajuato, GTO, Mexico

**Mario Pedraza-Reyes**
Departamento de Biología, División de Ciencias Naturales y Exactas, Universidad de Guanajuato, Noria Alta S/N, 36050 Guanajuato, GTO, Mexico

**Iris Heidmann**
Enza Zaden, Research and Development B.V. 1600AA Enkhuizen, The Netherlands

**Grit Schade-Kampmann**
Amphasys AG, Technopark Lucerne, 6039 Root D4, Switzerland

**Joep Lambalk**
Enza Zaden, Research and Development B.V. 1600AA Enkhuizen, The Netherlands

**Marcel Ottiger**
Amphasys AG, Technopark Lucerne, 6039 Root D4, Switzerland

**Marco Di Berardino**
Amphasys AG, Technopark Lucerne, 6039 Root D4, Switzerland

**Kieran O'Neill**
Terry Fox Laboratory, BC Cancer Agency, Vancouver, British Columbia, Canada
Bioinformatics Graduate Program, University of British Columbia, Vancouver, British Columbia, Canada

**Nima Aghaeepour**
Terry Fox Laboratory, BC Cancer Agency, Vancouver, British Columbia, Canada
Bioinformatics Graduate Program, University of British Columbia, Vancouver, British Columbia, Canada

**Josef Špidlen**
Terry Fox Laboratory, BC Cancer Agency, Vancouver, British Columbia, Canada

**Ryan Brinkman**
Terry Fox Laboratory, BC Cancer Agency, Vancouver, British Columbia, Canada
Department of Medical Genetics, University of British Columbia, Vancouver, British Columbia, Canada

# LIST OF ABBREVIATIONS

| | |
|---|---|
| ALLs | Acute lymphoblastic leukemias |
| ABA | Allen Brian Atlas |
| AC | Alternating current |
| ACS | Archival Cytometry Standard |
| CDPs | Cell division proteins |
| CPS | Chloroquine chemoprophylaxis |
| CLR | Classification Results |
| CD | Cluster of differentiation |
| EB | Ethidium bromide |
| EEF | Exoerythrocytic form |
| EQA | External quality assessment |
| FCM | Flow cytometry |
| FCS | Flow Cytometry Standard |
| FITC | Fluorescein isothiocyanate |
| FRET | Fluorescence resonance energy transfer |
| FACS | Fluorescence-Activated Cell Sorting |
| FACS | Fluorescent activated cell sorters |
| FISH | Fluorescent in situ hybridization |
| FVFC | Functional virtual flow cytometry |
| GCNA | Gene coexpression network analysis |
| GEO | Gene Expression Omnibus |
| GFP | Green fluorescent protein |
| HIV | Human immunodeficiency virus |
| IFC | Impedance flow cytometry |
| IQC | Internal quality control |
| ISAC | International Society for Advancement of Cytometry |

| | |
|---|---|
| MFI | Mean fluorescence intensity |
| MIC | Minimal inhibitory concentration |
| MRD | Minimal residual disease |
| MST | Minimum spanning tree |
| MDR | Multiple drug resistance |
| NND | Nearest neighbor distance |
| NA | nucleic acid |
| PNH | Paroxysmal nocturnal hemoglobinuria |
| PMT | Photomultiplier tube |
| qPCR | quantitative polymerase chain reaction |
| RAS | Radiation attenuated |
| SPICE | Simplified Presentation of Incredibly Complex Evaluations |
| VEC | Vaginal epithelial cells |
| VBNC | Viable but nonculturable cells |
| WHO | World Health Organization |

# PREFACE

How would it be to have a technique that aids in the analysis of several parameters at the same time? The technique of flow cytometry has its origins in early stages of the Second World War and today stands as a much used analytical and diagnostic tool. The term "cyto" refers to cell and flow refers to cells in a suspension that are measured on the basis of their property such as size or granularity or the expression of a particular protein or sequence.

The field of flow cytometry is much employed in research to scan for sub-populations of cells. Its use in hematology or blood studies has been invaluable in looking at various cells of the blood system. It has been used to identify/diagnose microbes or diseases caused by them making it a valuable tool in a clinical scenario. Whether it is an immunological disease or a serious one such as cancer, this technique has proved its mettle.

Initial instruments used were huge and bulky and required manual analysis: this scenario has changed nowadays with benchtop systems that use automation and computers to analyse. Flow cytometers are also used for sorting of cells: based on a fluorescent tag the machine can sort cells based on antigens or populations.

The technique has found application in expression of proteins, cell cycle studies, RNA analysis as well as analysis of a population of cells in a mixed population. This book covers the principles and working of flow cytometry along with chapters explaining the role of these instruments in microbiology, research and a clinical setting. Several case studies and figures have been added to enable a reader to grasp the concepts quickly. The editors hope this book will be a must have for anyone who wants to get the concepts in a nutshell.

# CHAPTER 1

# PRINCIPLES OF FLOW CYTOMETRY

**Shiv Sanjeevi Sripathi**

Shiv Sanjeevi Sripathi secured second rank in Mumbai University and first rank in his college at his undergrad level and in his college during his Masters and has worked on a stem cell project at the Specialized Centre for Cell Based Therapy (SCCT), KEM Hospital at Mumbai, following which he was a Junior Research Fellow at Microbiology & Cell Biology Department, Indian Institute of Science, Bangalore from August 2009 to November 2011. The field of work involved cloning of cell wall genes and transcription factors in E. coli & M. smegmatis. This was followed by a stint as a Research and Development Executive at Transasia Biomedicals, Mumbai where he worked on patent writing and Immunoassays from 2011 till 2013. In 2013, he entered the teaching field where he has been tutoring biology and chemistry to the students of 11th and 12th grades along with English grammar and communication skills to people of all age groups. Shiv also loves helping people write scientific manuscripts and has co-edited several books on life sciences and aspects of biotechnology with Apple Academic Press.

## 1.1 INTRODUCTION

The ability to measure several characters of a cell at the same time is addressed by the technique of flow cytometry. Using several instruments, features of a cell are studied in a suspension state and system. A definition of the system is "measuring several characters of a cell in suspension at the same time utilizing a measuring device."

**Table 1.1:** Applications of flow cytometry

| **Field** | **Use of flow cytometry** |
|---|---|
| Oncology | DNA content or looking at market |
| Immunology | • For transplants<br>• Histocompatibility matching |
| Hematology | • Counting of cells<br>• Identification of phenotypes of cells<br>• Analysis of auto immune disorders<br>• Diagnosis |

The size of cells was first measured as a parameter in the first flow cytometer. There are later advancements that can analyze several parameters at the same time such as the granular nature of cells apart from size and granularity along with fluorescence.

Flow cytometry is of two types:

- *Non-sorting:*
  This system can scatter light and emit fluorescence.
- *Sorting:*
  This system can scatter as well as separate cells on the basis of properties.

The flow cytometer are based on fluidics; this directs the cells or particles in solution that is exposed to light; upon excitation and light scatter from the particles the light signals are read as computer signals. This generated data on the computer is proportional to the intensity of light.

(Wilkerson, 2012).

Coulter counter (Beckman Coulter **M**) was one of the first instruments to be used. The counter measures the resistance of a cell in electric current as it traverses a small hole (an orifice) that is indicative of its numbers or size. A combination of fluorescence signals coupled with excitation with laser is reflective of flow cytometry. The extent of scatter of light can reflect the granularity and size of a cell. The use of stains or antibodies can allow for the detection of specific nucleic acids or proteins expressed on a cell.

The system can be akin to a fluorescence microscope. Here, a microscope uses a lamp as light source that is focused by an objective and detected by our eye. With the use of stains different size of cells can be viewed by an observer. In flow cytometry instrument, while a laser is the source of light that is focused by an objective, a photomultiplier tube functions as a detector.

**Table 1.2:** Microscopy vs. Flow cytometry

| **Feature** | **Microscopy** | **Flow cytometry** |
|---|---|---|
| Analysis | **100 cells/min** | **50-4,000 cells/s** |
| Use | Analyze morphology or look at location, study structure, study of sections such as that of tissues, Distribution of a probe within a cell | This system doesn't permit the study of an intracellular probe. |
| Numbers | Fewer: less than or equal to 1000 | Higher rates of analysis |
| Requirement | | Single cell suspension |
| Added advantage | | Multiparameter analysis |

While the binding of a specific probe emits a particular wavelength, detection is permitted in a fluorescent microscope. It develops a focused image. This is in contrast to a flow cytometer where light is detected and converted into a signal for analysis (Bakke, 2001).

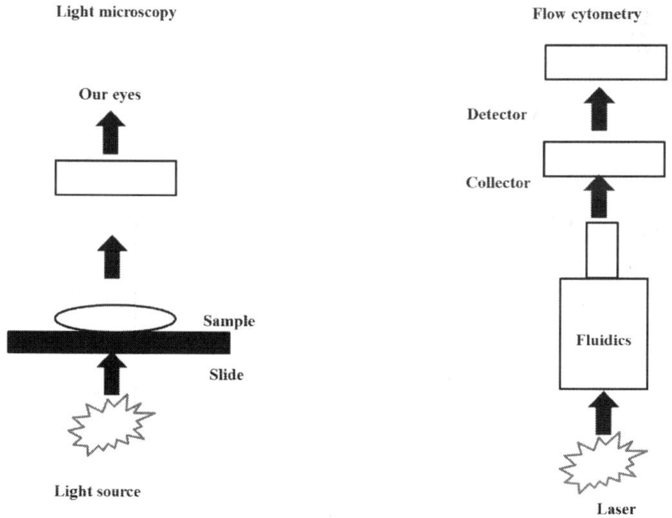


**Figure: 1.1:** Microscopy vs. Flow cytometry.

## 1.2 COMPONENTS OF A FLOW CYTOMETER

The optical and fluorescence characters of single cells that include microbes, nuclei or beads or chromosomes are measured. Several properties such as forward angle light scatter as size; right-angle scatter that represents internal complexity. The binding of fluorescent dyes with cell components or intercalation with nucleic acids and subsequent exposure to light causes their excitation to higher energy levels. A return to their original energy level causes emission of light; different properties can be studied using different colors or wavelengths of light. The use of dyes in tandem can broaden the spectrum or scope of study.

There is an isotonic fluid as a sheath that surrounds cells in suspension in order to draw the cells into a stream. There is an "interrogation point" where cells pass through one by one. A source of monochromatic light such as laser illuminates the cells; the light that is emitted is made to pass through several filters and mirrors. The light signals are isolated by this system of optics aid in selecting a particular wavelength. This is followed by digitization and analysis by a computer: the data can be displayed as 2D dot plots or histograms.

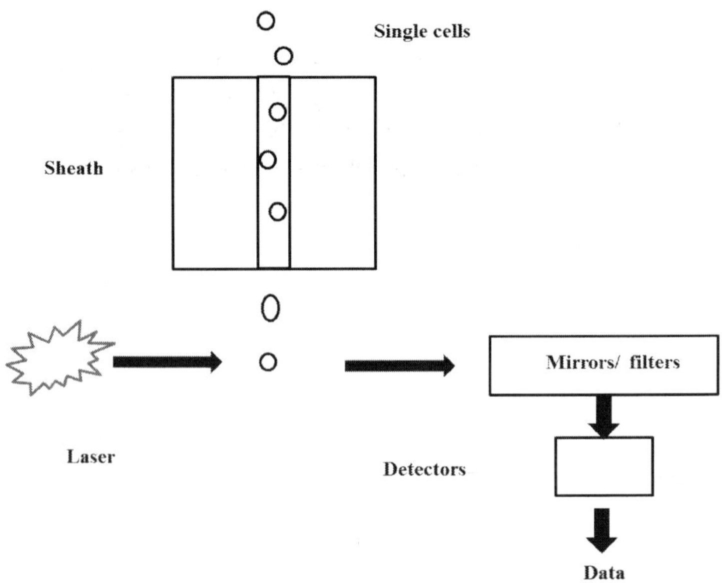


**Figure 1.2:** Schematic of flow cytometry.

(Brown and Wittwer, 2000)

## 1.3 HOW DOES FLUORESCENCE WORK?

Light energy is absorbed by a molecule that shows fluorescence can absorb light energy that makes the molecule excited that refers to an electron rising from a lower level to a higher level. When this electron returns to the lower level the surplus energy is given out as photon known as fluorescence (figure 1.3).

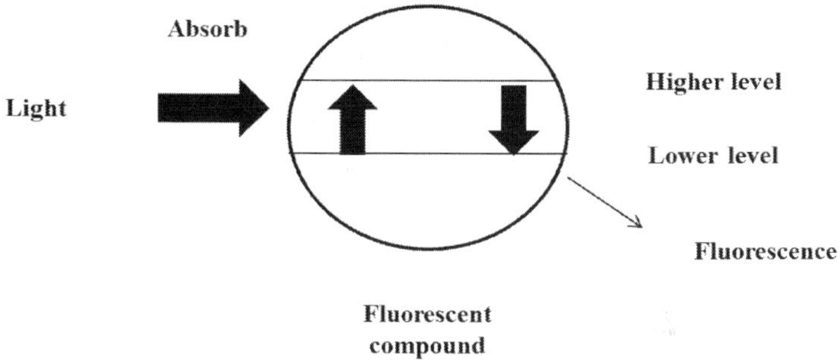


**Figure 1.3:** Fluorescence.

(Ormerod, 2009; Adan et al. 2015)

A fluorochrome is defined by the excitation wavelength called absorption spectrum and the wavelength at which it emits light called emission spectrum. The value of the latter would be more than the former. The function of a fluorochrome is defined using Stoke's shift (figure 1.4) that is the difference between the 2 spectra.

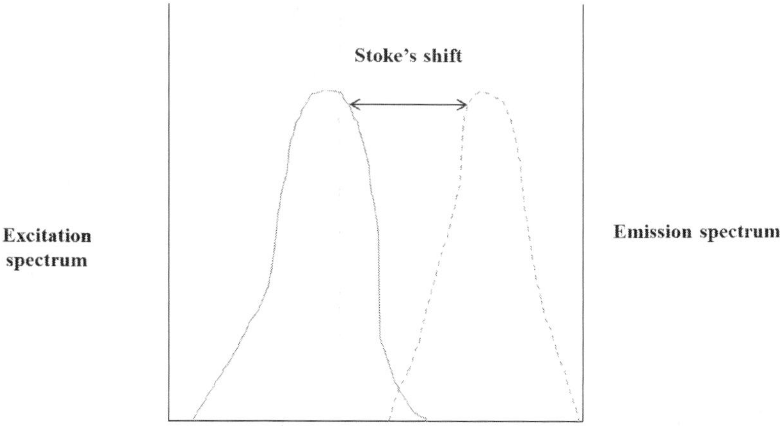


**Figure 1.4:** Stoke's shift.

With larger values of Stoke's shift, a larger separation is achieved between excitation and emission leading to resolution using filters. Dyes such as fluorescein isothiocyanate (FITC) show a range of absorption at 400-550 nm with a highest at 490nm (Ormerod, 2009; Adan et al. 2016).

## *Flourochromes in flow cytometry*

For proteins, antibodies serve as their probes. They are labelled by fluorescein isothiocyanate (FITC) or allophycocyanin (APC) or phycoerythrin (PE). The choice would depend on the laser source; the primary choice would be FITC followed by PE. Other than antibodies, cDNA or lectins, avidin or hormones can also be labelled with such fluorochromes.

FITC can form complexes with fair stability with lysine molecules on a protein. With a maximum value of absorbance at 490nm it can be used for staining single colors. Challenges with this system include bleaching upon light exposure and sensitivity to pH. This has led to several modifications such as Alexa™ that offer advantages such as stability (Macey, 2010).

Pigments based on photosynthesis include PE and APC that offer high level of fluorescence. While the former needs argon laser the latter with higher absorption requires the use of helium–neon laser. Challenges with these systems are stearic changes due to change in structure as well as high background if inefficient washing is done (Telford, 2015).

For nucleic acids, fluorochromes are also used for analysis of cell cycle or looking at measurements of DNA. Several dyes include:

*Hoechst and DAPI that bind to AT rich regions.

*7-Aminoactinomycin D (7-ADD) mostly stains dead cells and binds to GC regions; antibodies coupled to FITC- and PE can be used.

* Green fluorescent protein (GFP) is used for sorting transfected cells.

* JC-1 is seen in mitochondria of healthy cells and can hence be used for studying potential.

Another technique involves the use of two fluorochromes: this can cause a single wavelength to excite several colors using fluorescence resonance energy transfer (FRET). The excitation of the first dye leads to its energy being transferred to the second to cause emission (Reviewed by Adan et al. 2016).

## 1.4 PRINCIPLES UNDERLYING PORTIONS OF FLOW CYTOMETRY

### 1.4.1 Fluidics

The cells are transported via the instrument from their suspension to be analyzed. There is the specimen in question along with a sheath fluid (a buffered saline). The sheath fluid is applied into the flow chamber. In order to facilitate the cells to flow along a single line, the sheath fluid pressure is lesser than sample pressure. This principle of hydrodynamic focusing aids in the illumination of cells in a uniform fashion. The rate of injection is dependent on the experiment; slow rates are applied for high resolution applications (DNA) while qualitative analysis can use fast flow rates. In order to achieve good analysis, bubbles should be avoided while the solution should be uniform (Bakke, 2001; Adan et al. 2016).

### 1.4.2 Optics

A fluorescent tag attached is excited by the illumination of light mostly laser. The source of light is commonly is blue argon laser as well as red helium-neon laser or mercury arc lamp. The light is passed through optics to form a beam as an ellipse that is high in intensity in the middle. It is focused on the flow chamber where the sample is also injected that leads to light hitting the sample (Bakke, 2001).

In one pattern, the diffraction of light around cell edges causes a diffraction pattern in the path of the laser. The scattered light can be equated with cell circumference with the wavelength of incident light. This is *forward scatter* or also called *forward angle light scatter* shortly called FALS, where the emission is collected on the same axis as the incident laser.

Another system involves collection of the scatter 90 degrees to the incident laser. This is *side scatter*, or *right-angle light scatter* referred to as RALS, or *90- degree light scatter* (90LS). It is based on light that is reflected from internal structures and is reflective of cell granularity.

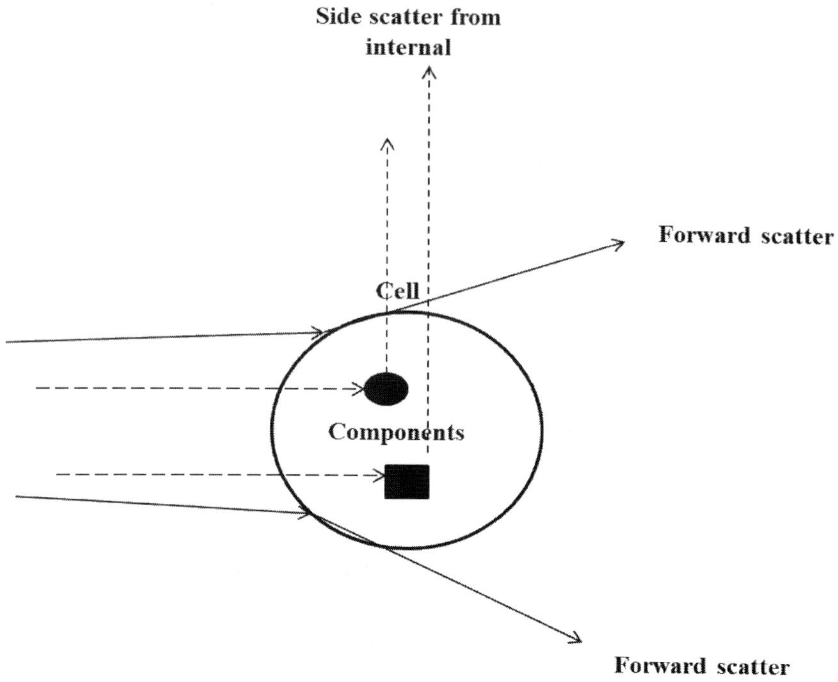


**Figure 1.5:** Forward and side scatter.

The collection of light is done by collection optics to collect light from the interaction of the laser and sample along with mirrors/ filters that channelize light of appropriate wavelength. Based on the particular fluorescent dye several filters include:

*Band pass filters are for a small range of wavelengths that is near the value of emission of the dye.
*Short pass filters are for wavelengths of that are less than or equal to required wavelengths.
*Long pass filters are for wavelengths more than or equal to a specified wavelength.

### 1.4.3 Electronics

The generated light signals are later processed into electronic ones using photodetectors. There are 2 type systems called photodiodes (PDs) and photomultiplier tubes (PMTs). The latter show an enhanced sensitivity due to increased efficiency of conversion. These generated voltage systems require conversion to a digital format as they are in analog form.

The initial signal is amplified in two ways:

- Linear amplification increases in a range of 200-1000 and is used for direct relationships and scatter.
- Logarithmic amplification amplifies to a range of 2,000- to 10,000 and is applied for immunofluorescence studies.

(Bakke, 2001; Adan et al. 2016).

## 1.5 PROCESSING OF DATA

The light signal generated is transformed into electric currents that are subjected to amplification. This analog signal undergoes conversion into digital one for input into a computer. This digital data is expressed as:

1 parameter histogram reflects average value of fluorescence in a study. In the case of logarithmically amplified signals the mean geometric fluorescence is applied.

2 parameter plots have 4 quadrants that have fractions or percentages of total cells. Based on the fluorescent and non-fluorescent cells the difference in expression can be studied.

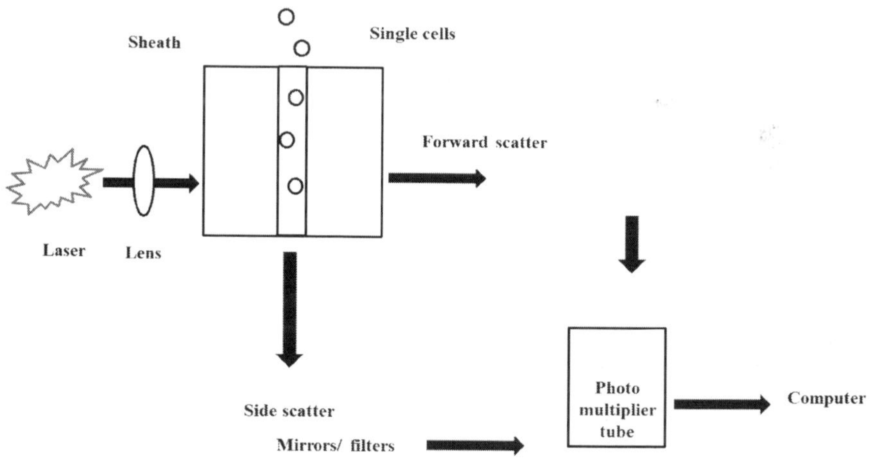


**Figure 1.6:** Data processing in flow cytometry.

## 1.6 GATING

This refers to define a subset of desired data; the use of a graphical or numerical boundary to specify a population of cells or particles is a "gate". Consider an example of a blood sample where the cells of interest are lymphocytes. A gate is set up using properties such as cell size or an antibody that binds to it. Several types of plots can show the data such as histograms or 2D dot plots or density plots, contour plots or tomogram plots that is 3D (Adan et al. 2015)

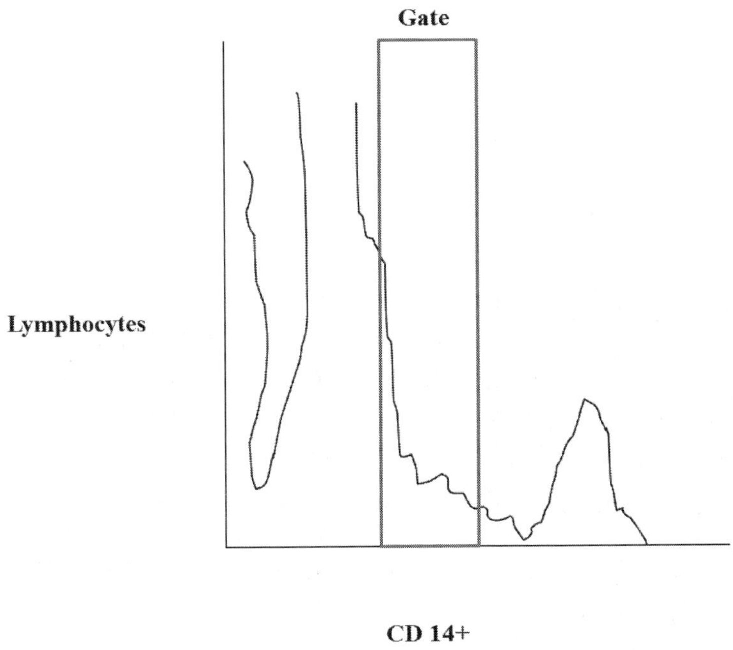


**Figure 1.7:** Gating for cells (here, lymphocytes).

## 1.7 SORTING OF CELLS

Flow cytometry is also applied in the realm of sorting of cells. Cells whose features have been defined are "caught" and separated that are then analysed for structural and functional purposes. Cell sorters are applied for this purpose. The sorters operate on the principle of charged droplets subjected to electrostatic deflection.

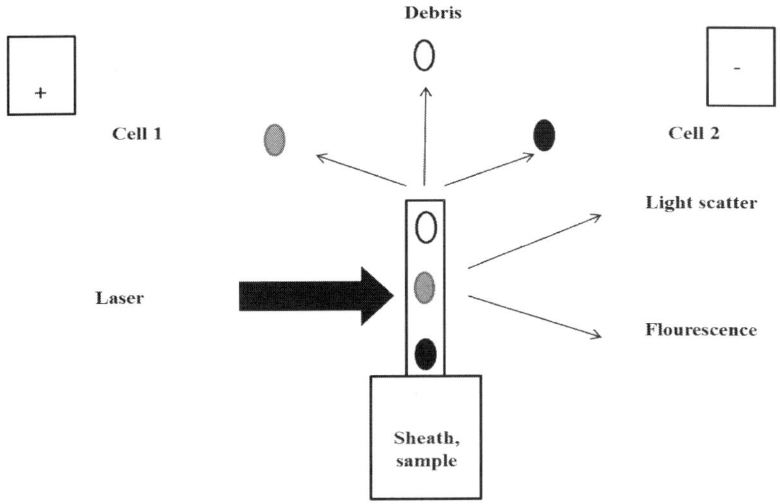


**Figure 1.8:** Flow cytometry for cell sorting.

The system involves injection of cells via a nozzle to create a stream of droplets. Following passage through laser, the droplets are given a charge using an electrode. Based on charges of the droplets, negatively charged particles are deflected towards a positively charged plate while positively charged droplets are deflected using a negative electrode while and the ones with no charge are collected as debris (Davies, 2010; Adan et al. 2016).

There are Fluorescent activated cell sorters (FACS) that can separate cell mixtures using labelled cells using fluorescent staining (Wilkerson, 2012). Based on the tag attached and properties, cells under study are selected as they are in passage via the laser. Such tagged cells are charged and separted into different collection systems (Davies, 2010; Adan et al. 2016).

## 1.8 QUALITY CONTROL

In order to achieve optimum results in a flow cytometry experiment, quality control is vital. There are the following systems of quality control.

- The machine should be subjected to evaluation where standardized fluorescent beads are used. Such beads are stable and can be used to plot data and assess the functioning of the machine. The instrument can be analyzed using these reproducible systems.

- In another level of control, material is tested for the staining and instrument settings. The test materials used are close copies of the real samples.
- Here control materials are used (such materials are available on a commercial level). They can be used to measure instrument sensitivity (Bakke, 2001).

## 1.9 A NEW PERSPECTIVE

The path of flow cytometers have seen a progress from their initial days where laser excited fluorescein and was found to augment sensitivity. Today we use digital platforms that combine these signals with a digital format and then aid in the use of fluorescent probes for analysis of cell surfaces and components such as nucleic acids and proteins. The analysis of mixtures for several samples has been facilitated by this system.

Let us look at several milestones over the years using the following table: Adapted from Moore and Yvon, 2012)

**Table 1.3:** Milestones

| **Property** | **1900s** | **2012** |
|---|---|---|
| Preparation of samples | Fluorochrome used was FITC | • Fluorochromes have been developed that allow analysis of several wave-lengths (Polychromatic analysis)<br>• Several tandem dyes and quantum dots have also been reported. |
| Optics | Lasers were water cooled and were of lower sensi-tivity and lesser excitation wavelengths | • Solid state diode lasers are sued that have enhanced sensitivity and more excitation wavelengths. This allows the detection of several fluorochromes.<br>• The samples processes have higher throughput due to multiple wavelength analysis with lower sample volumes. |
| Throughput | 1,000s events/second | 100,000s event/second |
| Signals | Analog | Digital that has led to increased linearity. |
| Analysis of data | It was manual and 2 dimensional that made it laborious and subjective | The data analysis has become automated based on algorithms that has led to increase in throughput as well as study of complex samples with lower bias. |
| Size | Larger | Benchtops suitable for use in clinics |

# 1.10 CONCLUSION

One of the most powerful techniques to analyze samples in a short span of time is how researchers describe the field of flow cytometry (Adan et al. 2016). This rapid analysis is the need of the hour while current market trends look at analysis of single cells. The field of genomics, proteomics and metabolomics are seeing further fine tuning with study of single cells or cytomics.

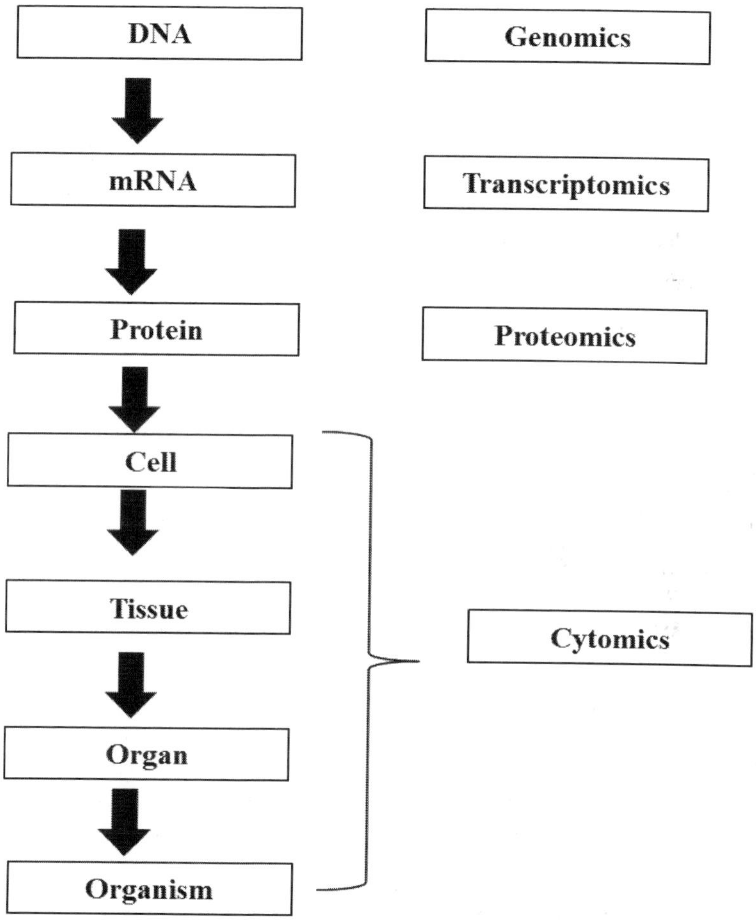


**Figure 1.9:** Emerging fields.

A 2012 report from HTStec Ltd concluded based on responses of 202 participants that specificity is higher with the use of single cells rather than population studies (Single cell, 2012).

**There are several feathers in the cap of the field of flow cytometry:**

It can detect and aid in the counting of populations and sub-populations of cells in a mix of cells (take for example blood).

Several biomarkers have been identified as a particular phenotype had to be separated from a mix that can see applications in the clinic and research.

There is the field of cytomic profiling that is the union of high dimensional flow cytometry and systems biology studies. Such a field can serve as a tool for diagnosis, prognosis and therapy! (Moore and Yvon, 2012)

The 21st century has seen mergers of technology along with development of flow cytometry. The analysis of single cells has been facilitated with such cytometry. Such approaches can find a huge application in today's much needed personalized medicine (Moore and Yvon, 2012).

A continual thrust in research of flow cytometry for further improvements is expected that can serve several fields of research, medicine and science (Adan et al. 2016).

## 1.11 REFERENCES


1. Adan A, Alizada G, Kiraz Y, Baran Y, Nalbant A. (2016). Flow cytometry: basic principles and applications, Crit Rev Biotechnol, doi: 10.3109/07388551.2015.1128876
2. Bakke AC. (2001) The Principles of Flow Cytometry. Lab Med, 4 (32).
3. Brown M, Wittwer C. (2000). Flow Cytometry: Principles and Clinical Applications in Hematology. Clin Chem, 46:8(B) 1221–1229.
4. Davies D. (2010). Cell sorting by flow cytometry. In: Macey MG, ed. Flow cytometry: principles and applications. Totowa (NJ): Humana Press, 257–76.
5. Macey MG. (2010) Principles of flow cytometry. Flow cytometry: principles and applications In: Macey MG, ed. Totowa (NJ): Humana Press, 1–15
6. Moore J, Yvon P. (2012). High dimensional flow cytometry comes of age. European Pharmaceutical Review, 17(4), 20-24.
7. Ormerod MG. (2009) Flow cytometry: A basic introduction. Chapter 3: Fluorescence and fluorochromes. ISBN 978-0-9559812-0-3.
8. Single Cell Technologies Trends. (2012) HTCtec Ltd Report, June 2012
9. Telford WG, Shcherbakova DM, Buschke D, et al. (2015). Multiparametric flow cytometry using near-infrared fluorescent proteins engineered from bacterial phytochromes. PLoS One, 10, e0122342.
10. Wilkerson MJ. (2012). Principles and applications of flow cytometry and cell sorting in companion animal medicine. Vet Clin North Am Small Anim Pract, 42, 53–71.

# CHAPTER 2

# FLOW CYTOMETRY IN THE REALM OF BACTERIA

**Shiv Sanjeevi Sripathi**

Shiv Sanjeevi Sripathi secured second rank in Mumbai University and first rank in his college at his undergrad level and in his college during his Masters and has worked on a stem cell project at the Specialized Centre for Cell Based Therapy (SCCT), KEM Hospital at Mumbai, following which he was a Junior Research Fellow at Microbiology & Cell Biology Department, Indian Institute of Science, Bangalore from August 2009 to November 2011. The field of work involved cloning of cell wall genes and transcription factors in E. coli & M. smegmatis. This was followed by a stint as a Research and Development Executive at Transasia Biomedicals, Mumbai where he worked on patent writing and Immunoassays from 2011 till 2013. In 2013, he entered the teaching field where he has been tutoring biology and chemistry to the students of 11th and 12th grades along with English grammar and communication skills to people of all age groups. Shiv also loves helping people write scientific manuscripts and has co-edited several books on life sciences and aspects of biotechnology with Apple Academic Press.

## 2.1 INTRODUCTION

Flow cytometry (FC) is a well-accepted laser-based technique which analyzes the traits of cells or particles. It divulges the expression of the intracellular molecules, cell surfaces, thus defining the various types of cells in a pool of diverse cell population. It helps in assessing the purity of isolated subpopulations and studying the volume and size of cells. It also aids in analyzing the different facets of single cells. Though FC came into view 1960's as a technique for investigating blood cells, advanced studies in flow, electronics, optics and probing mechanisms lead to using FC in various studies intending knowledge in diverse biological systems. FC in addition with FC based cell categorization have been recognized and used as essential tools in recent cell and developmental biology where the technique is utilized for cell counting, biomarker detection, cell sorting and protein engineering.

At large, it is used for measuring the intensity of fluorescence exhibited by antibodies or particles labeled with fluorescent dyes, helping in the detection of proteins or ligands which attach to certain specific molecules of the cell for example, the propidium iodide bond with the DNA. The FC staining technique engages in establishing a single cell suspension from the given tissue samples or cell culture. The cells are incubated in microtitre plates or test tubes with fluorochrome labeled antibodies which are then studied on the flow cytometer. In a FC, the cells or particles are passed through an interrogation point which allows a laser beam to impact them and the light which the particles have absorbed gets scattered owing to its extrinsic or intrinsic physical properties, which are then measured (Shapiro, 2000).

FC is regularly used for the identification of health disorders, especially blood cancers. It is also used extensively in clinical practices, basic research and also clinical trials. It aids in physically sorting and purifying the particles of interest on the basis of their optical properties. It is called as cell sorting and the apparatus is referred to as the cell sorter. Recent FCs are capable of investigating several thousand particles per second and the cell sorters can detach and segregate particles at similar rates that have specific optical properties. It is similar to the microscope but instead of producing the image of the cell or particle, FC offers high throughput and offers automated quantification of the specific optical factors on cell to cell basis on a large scale.

## 2.2 APPARATUS

The FC comprises of fluidics, electronic systems and optics. The fluidic system allows the respective cell or particle in a fluid stream to one by one, at high speed, go through a laser beam in the interrogation point which is in the centre of the centre of the core stream. Here, the diameter of the particle's path is directed by the laminar flow applied by the adjoining sheath stream, allowing only one cell to pass at a time.

The optic system comprises of beam splitters, lenses and filters that direct incident light rays emitted by cells or the fluorescent light due to cell illumination, to the light detectors. The optics system comprises of a light source, commonly used is blue Argon ion laser which is at 488nm. This beam focuses on the cell present at the interrogation point by means of lens. At the laser intercept, single or multiple laser beams light up the cells so as to evaluate their light scattering characteristics while exciting the fluorescent particles/molecules.

Appropriate lasers can stimulate the naturally fluorescent molecules or particles under study and also fluorescent molecules exclusively attached to probe (stain) or tag the various components of the cells. The cell or particle under investigation then scatters light in all angles of which light dispersed at acute angles, also termed as forward scatter indicates the size of the particle. Light scattered at 90°/wide angle is called as side scatter. It is relative to the complexity or roughness of the particle. Side scattered light and fluorescence received, accurately made parallel and aimed at 90° throughout the pathway connecting a sequence of beam splitters or dichroic mirrors which allow certain redundant wavelength of light while reflecting the essential light through a band pass filter (green, yellow, orange and red light), finally to photomultiplier tube (PMT)

Forward scattered light is identified ahead of the incident laser beam route either by a PMT or a photodiode. PMT amasses photons and forces out electrons while amplifying the signal as PMT can generate a few hundred electrons in support of every photon. This result in the generation of concise current pulses which get amplified changed to voltage and transformed to figures by analog signal processing electronics. In FC, all the events generated per second can be represented and results obtained show a series of plots and figures to be interpreted. Thus, the light signals obtained by detectors get transformed into electronic signals by the electronic system.

It filters and develops computer based algorithms. The sample's characteristics such as its relative size, internal complexity, relative granularity and relative intensity of fluorescence of the molecule under examination are measured. Thereafter, fluorescence activated cell sorting (FACS) or complementary analysis (eg via microscopy) will aid in further research. Fluorescent dyes may be implied indirectly (by using antibodies) or directly to label the various cellular apparatus like DNA, specific nucleic acids (NAs), intracellular enzymes and structural proteins, surface proteins and receptors, membrane properties, secreted proteins, ion fluxes or small molecules and organelles. Not all of these have been employed in the study of bacteria but can be used in future research (Ambriz-Aviña, et al., 2014).

Light scattering is victorious in determining the individual characteristics in bacteria which includes accumulation of poli ($\beta$-hydroxybutyrate) and also cell death and filamentation (Walberg, et al., 1997). Light scattering reveals the intrinsic properties of cells, however, it is complicated when it comes to the study of the various bacterial species and to aid this process, the cells should be marked with fluorescent dyes (Veal, et al., 2000). Light scattering has triumphed for categorization and examining the finest clones in an *E. coli* culture proficient of generating the growth hormone or interferon (Wittrup, et al., 1988). The ability of conventional FCs for measuring the extrinsic or intrinsic fluorescence in cells acts as a trait for assessing protein identity, cell viability and enzymatic activity (Steen, 1990).

## 2.3 APPLICATION OF FC FOR BACTERIAL ANALYSIS

### 2.3.1 Responses

Advanced research on FC accompanied with novel invention of cellular reporter probes aids in applying the technique for analyzing the bacterial responses at physiological levels. Research by Ambriz-Aviña, et al., allows FC to illustrate diverse physiological conditions present in bacteria along with inclusion of its reaction with antibiotics, physical and cytotoxic chemical features, cell differentiation through the formation of biofilm, interactions between the host and pathogen and the processes leading to progressive pathways like sporulation. As FC can be employed for single-cell level studies, their research reveals the strength of the technique for granting precious information in regards to the heterogeneous allocation of specialized and differently responding cells along with imminent cell interactions, like in the formation of biofilms (Ambriz-Aviña, et al., 2014).

### 2.3.2 Study of heterogenous populations

Recent studies prove that simple laboratory cultures not only comprise of isogenic bacteria but are composed of heterogeneous subpopulations which react diversely to environmental changes (de Jong, et al., 2012). Valuable information is procured while studying the batch cultures, however, cell to cell interactions and intracellular heterogeneity present in isogenic cultures of bacteria, should be considered. Thus, studies on single-cell analysis united with subpopulation analysis techniques aid in their detailed studies in laboratory cultures. FC is thus employed for all these detailed studies in bacterial studies.

Morphogenesis and cell differentiation result in heterogeneity in population where minimum culture sample or changes in environmental changes may lead to huge changes occurring at the differentiation stage of particular individuals in the population. Studies in bacilli sporulation (differentiation causing maturation of the endospore) has been accepted as a bistable technique which has the co-existence of two diverse populations of cells; sporulating and non-sporulating bacilli.

This bistability can be clarified by the individual levels of Spo0A (master sporulation factor) activation due to phosphorylation stages. It is observed that in clostridia, this sporulation process is not as much synchronous and is variable compared to bacilli. This is evident from FC analytical studies on *Clostridium acetobutylicum* sporulation. It is a multi-stability stage in microbial systems is defined as instantaneous and meta-stable coexistence of more than two diverse cellular phenotypes coming up from clone populations.

Equivalent heterogeneity in populations can be noticed in other prokaryotes like in streptomycetes and actinomycetes and also in fungal systems. As the production of metabolites is frequently differentiated or dependent on the morphogenic stage, these heterogeneous populations can lead to unpredictable metabolic outcomes which have many fundamental and practical applications. Morphogenesis and cell differentiation along with associated heterogeneity in populations; both in eukaryotic (as in fungus) and prokaryotic systems (as seen in clostridia, bacilli and streptomycetes) severely impact the development and production of primary and secondary metabolites.

Practical applications of heterogeneity observed in population plays a significant role in cases like industrial production of erythromycin. If there is lack of morphogenetic control or lack in accurate separation of *Saccharopolyspora erythraea*, it may result in ineffective fermentations. According to biotechnology, it is essential to see if this state be sustained autonomously, if the product be manufactured in a semi or continuous pattern and if the ineffective isolation in cells be discarded. This is where FC and or FACS come in handy to identify, study and re-clone useful cellular phases and phenotypes. It has been lately expressed in the identification of butanol formation in C. acetobutylicum. Results suggested that clostridial cell forms that are characterized by formation of granulose do not produce solvent and is rather a vegetative like cell type.

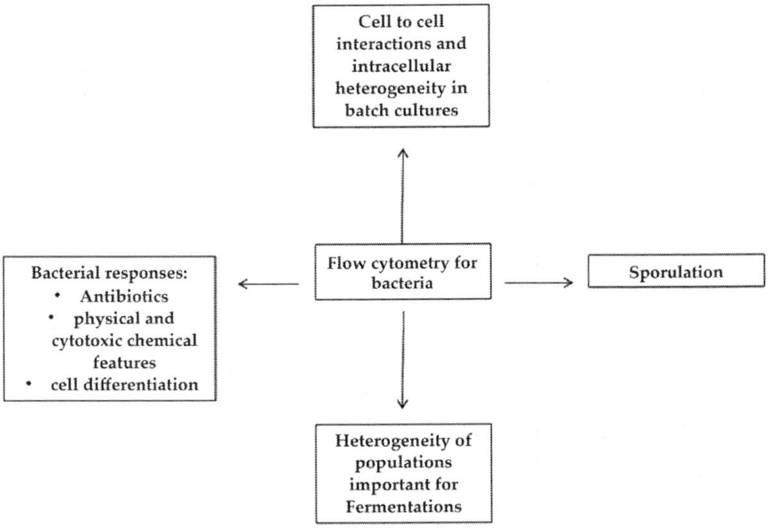


**Figure: 2.1:** Few applications of flow cytometry.

Lidstro and Konopka reviewed that in a bacterial culture; the response is on and off and depends on a collective response mechanism which elucidates a rise in physiologically diverse populations. This possibility that a diverse subpopulation may brunt the behavior of the whole population or the fate of the particular bacterial culture can depend on the temperament of response itself. For example, while a subpopulation manufactures and secretes a growth factor, an inhibitor or an autoinducer in the medium, it results in the response from all the remaining cells.

The occurrence of these heterogeneity responses may be due to the divergence in the environment which is caused by the outsized growth of bacterial cells while modifying the culture giving rise to intrinsic cellular noise along with extrinsic noise. This is based on the information that every cell has a diverse spatiotemporal distribution of various components of the cells i.e. the brink of molecules which switch on and off the gene expression, arises at unusual times in each cell as gene expression is stochastic (Ambriz-Aviña, et al., 2014). For the in-depth studies to measure the response of bacterial subpopulations and their physiological responses fluorescent techniques were found to be greatly essential and FC proves to be handy for accomplishing these results.

### 2.3.3 Cell sorting

FC can be outfitted with cell sorting ability for analysis by Fluorescence-Activated Cell Sorting (FACS). FACS is a registered brand by Becton-Dickinson Company but it is generally used generically in the scientific community. Various methods can be employed for confining of cells in sorting and the decisions may depend on one or several properties exhibited at the cellular level. Cells, be it a single cell or in millions, collected in tubes post sorting can later be employed for advanced assays, functional and molecular assays and microscopy.

FACS and FC can be applied from cell to progress in biology, development in medicine and also in the in depth study of microbial organizations. There are ample applications of FACS and FC for the detection and sorting of cells with advantageous properties for the fundamental and vital studies in cells, developmental biology and also in clinical studies (Perez and Nolan, 2006). These applications aids in the progression of FACS and or FC apparatus, fluorescent probes, technique for labeling cells, analyzing the fluorescent signals while sorting cells for further investigations. Of late, FC applications for investigations in bacterial systems have considerably developed highlighting its importance in microbial physiology, biotechnology and genetics (Tracy, et al., 2010).

Several studies are conducted on the nucleic acid (NA) content of each specific cell and cellular distribution in a subpopulation to find the NA containing components, to determine the membrane integrity of cells and to cross examine the proliferation stages of cells using FC. NA dyes are selective in staining RNA and DNA or both and can be selective for G-C verses A-T rich sequences along with their permeability to the cell membrane of bacteria.

The amalgamation of NA, intracellular pH, membrane potential and enzyme activity dyes may confer profound approach in studying the physiology of individual and or separate populations of bacterial cells. Multiparamteric FC (MPF) assays possess relevant steps for the identification and development of strains that exhibit complex phenotypic characteristics. It acts as a tool to assess the viability and improvise the discrimination of the physiological conditions of cells.

### 2.3.4 Study of phenotype

FC aids in measuring variety of phenotypic characteristics of particular microorganisms by evaluating an array of cellular characteristics. These characteristics can be labeled as intrinsic or extrinsic. Intrinsic traits can be measured directly without the need to label the cells and mostly comprises of Light Scattering (LS) characteristics of the cells. Forward Scatter (FSC, small angle scatter) seldom provides details about cell shape and size while Side Scatter (SSC, large angle scatter) usually depicts the internal granularity of cells. For measuring the extrinsic factors, fluorescence labeled probes or stains are implied to study cellular features and components like membrane potential, DNA content and the appearance of intracellular and surface proteins (Tracy, et al., 2010).

### 2.3.5 Expression of proteins

Non-invasive techniques used for labeling are used to evaluate the expression of proteins by FC. They are mostly based on protein tags which involve green fluorescent protein (GFP) or its derivative or reporter gene assays. Based on the genetic model used, expression of GFP suggests the transcriptional activity adopted by a promoter or state the expression levels regarding a particular protein (Tracy, et al., 2010).

### 2.3.6 Response to antibiotics

FC acts as a potential tool for analyzing the bacterial response exhibited towards antibiotic agents. With an emergence in resistant strains of bacteria, efforts are made to discover new antibacterial drugs (Negi, et al., 2008). This is best done with the employment of automated methods for assessing the amount of required candidates along with discovering the effective drugs to treat bacterial infections on bases of hours and not days (Shapiro, 2001).

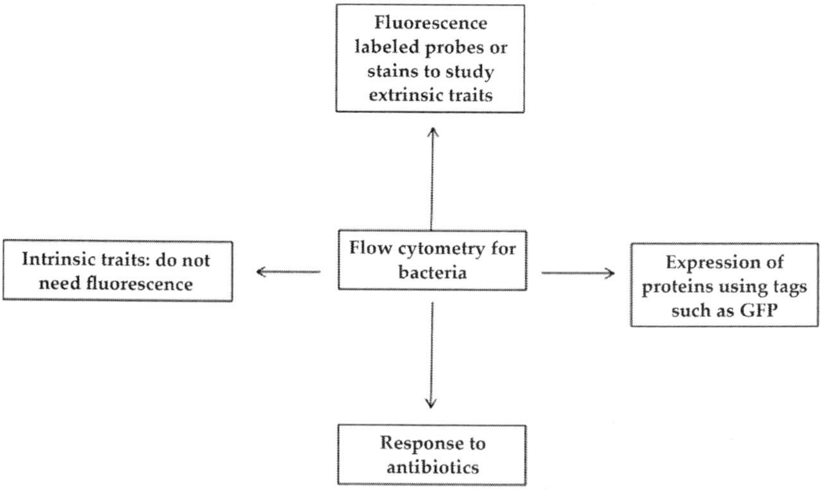


**Figure 2.2:** Flow cytometry and microbiology.

Researchers applied FC to investigate the vulnerability of growing *E. coli* cells to ampicillin and mecillinam. This was done by staining DNA along with a combination of ethidium bromide and mithramycin (Walberg, et al., 1996). Fluorescence was documented for measuring the DNA content where the forward scatter was indicative of cell size and the side scatter was indicative of protein content or the proportion of dry weight. These factors were then introduced and tested in cells exposed to antibiotics for half an hour and doubled at an hour and also tested with various drugs. Walberg's research suggested that FC when applied yielded the light scattering for studying the vulnerability of antibiotic drugs and also for the measurement of DNA content. He also suggested FC as an important tool for distinguishing cells and other components (eg debri), in clinical samples. Results obtained by microscopic interpretations were at par with the action of the tested drugs. In another study, Walberg, et al., documented the importance of the technique for investigation the drug response in a mixed culture of *E. coli* and *Klebesiella pneumoniae.* Both the microorganisms are related in the manifestation of urinary tract infections and both are susceptible to ampicillin. With the use of FC, the technique was useful in detecting the microbial infections by measuring the light scattering, cell number and the DNA content which was associated with fluorescence and all of this in an hour of incubation with ampicillin in the same sample (Walberg, et al., 1997).

### 2.3.7 Flow cytometry and food

Food borne diseases, both in developing and industrialized countries is at a high owing to the speedily evolving practices in food production. As per CDC estimates, 1 in 6 people gets affected by food poisoning in a year. Most of the diseases are as a result of the 31 causative pathogens that includes bacteria, parasites and viruses at large. FC has been used effectively for the detection of microbes involved in a particular contamination. It reduces the background fluorescence and improves the noise signal and thus measures the bacterial load whilst identifying the specific disease causing pathogens. Identification of bacterial culture is a long process which involves several procedures. The FC tool produces a result in a few minutes and if the sensitivity towards one cell in a whole complex subpopulation is required, it may take several hours. FC offers an effective approach in real time for assessing food safety which is of great importance in regulatory inspection and also in self- monitoring industry (Buzatu, et al., 2014).

## 2.4 CASE STUDY: ANTIMICROBIAL TREATMENTS AND FLOW CYTOMETRY

The use of several physical or chemical methods to target microbes can be attributed to several causes such as compromised cell wall synthesis, inhibition of DNA replication or translation etc. Such an understanding can aid in the mechanisms of resistance especially in today's scenario of medical or food microbiology. While the use of techniques such as metabolic analysis or plating or vital stains are efficient they involve labor and time. Another complication is wrong assumptions. For example, when bacteria are exposed to a stress and enumerated with classical plate count method, the live bacteria are counted. But are the other cells really dead? Reports point to the existence of called viable but nonculturable cells (VBNC) that are not counted in the test. This is where techniques such as multi-parameter flow cytometry (MP-FCM) make an entry (Léonard et al, 2016).

The use of appropriate fluorochromes can highlight the parameters of a cell: both from intrinsic and extrinsic details. Dyes such as SytoxTM or cyanines such as TO-PRO enter cells that have lost membrane integrity. Propidium iodide (PI) binds to dead cells to be used with green fluorochromes such as SYTO. Ethidium bromide (EB) can check for the pump system of a cell as it is pumped out if a cell has an intact pumping system but is retained by compromised cells. Living cells have a membrane potential of

100 to 200mV that is lost in the event of death as the concentrations of ions are different across cell membranes in a living cell. Viability can be seen with the use of dyes such as carbocyanins, DiOC*n*(3),or Rhodamine. While depolarized cells that are an outcome of treatment with antimicrobials can be seen with DiBAC4(3) (bis-(1,3-dibarbituricacid)-trimethineoxonol.

The integrity of membranes can be measured by the use of PI and Syto combination to differentiate between live, dead and injured cells in several studies. The activity of enzymes has also been measured to study anti-microbial effects. For example, carboxyfluorescein diacetate (cFDA) can indicate esterase activity that is indicative of anti-microbial application. This dye can diffuse across cell membrane and subjected to action by esterases to yield a fluorescent compound. Hence such fluorescent cells are indicative of viability.

Another approach to evaluate cells is the use of respiratory activity: 5-cyano-2, 3-ditolyl tetrazolium chloride (CTC) is reduced by the respiratory chain of bacteria to form a fluorescent compound.

The following table looks at few applications:

**Table 2.1:** Flow cytometry and anti microbials: few glimpses (Léonard et al, 2016)

| Dyes used | Application |
|---|---|
| DCFH-DA,2′,7′-dichlorofluorescein-diacetate that indicates reactive oxygen species (ROS) was used to evaluate | The dye was used to evaluate flow cytometry in the role of ROS and bacterial cell death.<br>The survival of staphylococcal species and *Escherichia* was reduced to between 60-70%. |
| • PI, Propidium Iodide;binds to nucleic acid<br>• DiBAC4, Bis-(1,3-DibutylBarbituricAC-id: record membrane potential<br>• CTC, 5-Cyano-2,3-ditolyl Tetrazolium Chloride;can indicate respiration | The effect of L-cysteine was shown as bacterio-static on *Listeria innocua* and *Escherichia coli* |
| • PI, Propidium Iodide;binds to nucleic acid<br>• DiBAC4, Bis-(1,3-DibutylBarbituricAC-id: record membrane potential<br>• DRAQ5, 1,5-bis{[2-(di-methylamino) ethyl]amino}-4, 8-dihydroxyanthra-cene-9,10-dione;stains DNA | Essential oil from *Coriandrum sativum* was shown to damage cytoplasmic membrane of *Candida* |

| | |
|---|---|
| • Sytox Green that measures membrane permeability<br>• DiBAC4, Bis-(1,3-DibutylBarbituricAC-id)TrimethineOxonol<br>• DiSC3, 3,3′-Dipropylthiadicarbocyanin-eIodide that measure membrane potential | The mechanism of death due to Scopolendin 2 in *E.coli* O157and *Candida albicans* was found to be pore formation in cell membrane. |

## 2.5 SUMMARY

FC is a mechanized technique, it saves time, is accurate and sensitive. As several millions of cells can be processed and studied at the same time, the statistics are enhanced. This results in the FC being an indispensible tool for the identification and quantification of cells present in low quantities in a given subpopulation. FC tool when employed for bacterial analysis, aids in recognizing and counting the bacteria, helps in determining the modifications in the cellular functions, metabolic activities and also in the identification of genes which are expresses specially when put through different conditions. FC supports the actual measurements for every cell that is assessed and in comparatively very less time with respective to other tools. Accurate amalgamation of dyes, damage as a result of antibiotics and any aggressive agents can be determined by using fluorescent post-enzymatic cleavage substrate analogs, which helps in determining the metabolic activity. When combined with cell sorting, FC aids in the separation of particular bacteria from subpopulations which display separate physiological traits compared to the entire culture. When such subpopulations are subjected to transcriptional or proteomic analysis, it becomes an influential tool for better understanding of the molecular and cellular mechanism which control the cell differentiation in bacteria.

## 2.6 REFERENCES


1. Ambriz-Aviña V, Jorge A. Contreras-Garduño, and Mario Pedraza-Reyes. Applications of Flow Cytometry to Characterize Bacterial Physiological Responses. BioMed Res Int. 2014: 1-14.
2. Buzatu DA, Moskal TJ, Williams AJ, Cooper WM, Mattes WB, Wilkes JG. An Integrated Flow Cytometry-Based System for Real-Time, High Sensitivity Bacterial Detection and Identification. PLOS ONE. 2014; 9(4):1-10.
3. de Jong G, Veening J, Kuipers OP. Single cell analysis of gene expression patterns during carbon starvation in *Bacillus subtilis* reveals large phenotypic variation. Environ Microbiol. 2012; 14(12): 3110-3121.
4. Léonard L, Bouarab Chibane L, Ouled Bouhedda B, Degraeve P and Oulahal N. Recent Advances on Multi-Parameter Flow Cytometry to Characterize Antimicrobial Treatments. Front. Microbiol. 2016; 7:1225. doi: 10.3389/fmicb.2016.01225
5. Lidstrom ME, Konopka MC. The role of physiological heterogeneity in microbial population behavior. Nat Chem Biol. 2010; 6(10): 705-712.
6. Negi PS, Jayaprakasha GK, Jena BS. Antibacterial activity of the extracts from the fruit rinds of *Garcinia cowa* and *Garcinia pedunculata* against food borne pathogens and spoilage bacteria. LWT: Food Sci Technol. 2008; 41(10): 1857-1861.
7. Perez OD, Nolan GP. Phospho-proteomic immune analysis by flow cytometry: from mechanism to translational medicine at the single-cell level. Immunol Rev. 2006, 210:208-228.
8. Shapiro HM. Microbial analysis at the single-cell level: tasks and techniques. J Microbiol Methods. 2000; 42(1):3–16.
9. Shapiro HM. Multiparameter flow cytometry of bacteria: implications for diagnostics and therapeutics. Cytometry. 2001; 43(3): 223-226.
10. Steen HB. Light scattering measurement in an arc lamp-based flow cytometer. Cytometry. 1990; 11(2): 223-230.
11. Tracy BP, Gaida SM, Papoutsakis ET. Flow cytometry for bacteria: enabling metabolic engineering, synthetic biology and the elucidation of complex phenotypes. Curr Opin Biotechnol. 2010; 21:85–99.
12. Veal DA, Deere D, Ferrari B, Piper J, Attfield PV. Fluorescence staining

and flow cytometry for monitoring microbial cells. J Imm Methods. 2000; 243(1-2):191-210.

13. Walberg M, Gaustad P, Steen HB. Rapid assessment of ceftazidime, ciprofloxacin, and gentamicin susceptibility in exponentially-growing *E. coli* cells by means of flow cytometry. Cytometry. 1997; 27(2):169-178.
14. Walberg M, Gaustad P, Steen HB. Rapid discrimination of bacterial species with different ampicillin susceptibility levels by means of flow cytometry. Cytometry. 1997; 29(3): 267-272.
15. Walberg M, Gaustad P, Steen HB. Rapid flow cytometric assessment of mecillinam and ampicillin bacterial susceptibility. J Antimicrob Chemother. 1996; 37(6): 1063-1075.
16. Wittrup KD, Mann MB, Fenton DM, Tsai LB, Bailey JE. Single-cell light scatter as a probe of refractile body formation in recombinant *Escherichia coli*. Nature Biotechnol. 1988; 6(4): 423-426.

# CHAPTER 3

# FLOW CYTOMETRY: APPLICATIONS

**Dr Prerna Pandey[1], Shiv Sanjeevi Sripathi[2]**

[1]Dr Prerna Pandey, a biotechnologist with several years of wet lab research experience, has worked as a senior research fellow at International Center for Genetic Engineering and Biotechnology, New Delhi. Her field of research in PhD included isolation and molecular characterization of Geminiviruses, genome sequencing, gene annotation, as well as gene silencing using the RNA interference technology. She has published papers in peer-reviewed journals in the field and has a submitted a number of annotated Geminiviral genome sequences in the GenBank, including two novel ones. Prerna has also completed her editing and proof-reading courses from SfEP, UK and now works as a freelance scientific editor, writer and a subject matter expert. She has edited and co-authored several books on a range of topics in Life-Sciences. When Prerna is not busy with her assignments, she loves visiting historical places.

[2]Shiv Sanjeevi Sripathi secured second rank in Mumbai University and first rank in his college at his undergrad level and in his college during his Masters and has worked on a stem cell project at the Specialized Centre for Cell Based Therapy (SCCT), KEM Hospital at Mumbai, following which he was a Junior Research Fellow at Microbiology & Cell Biology Department, Indian Institute of Science, Bangalore from August 2009 to November 2011. The field of work involved cloning of cell wall genes and transcription factors in E. coli & M. smegmatis. This was followed by a stint as a Research and Development Executive at Transasia Biomedicals, Mumbai where he worked on patent writing and Immunoassays from 2011 till 2013. In 2013, he entered the teaching field where he has been tutoring biology and chemistry to the students of 11th and 12th grades along with English grammar and communication skills to people of all age groups. Shiv also loves helping people write scientific manuscripts and has co-edited several books on life sciences and aspects of biotechnology with Apple Academic Press.

## 3.1 INTRODUCTION

With the advent of molecular biology techniques, it has been possible to achieve fast and sensitive diagnosis in cases that require urgent attention. One of the techniques is flow cytometry, relies on specific cellular parameters and their identification using certain fluorochromes that may be either independent or bound to specific oligonucleotides or antibodies. Flow cytometry has helped develop quantitative procedures to accurately assess antimicrobial susceptibility, drug cytotoxicity and in vitro antimicrobial activity, and most importantly, detect populations that respond differently to antimicrobial treatments. We make an attempt to provide an overview of the flow cytometric techniques, with some case studies that have applied this technology in biomedical research.

## 3.2 FLOW CYTOMETRY: PRINCIPLES

Flow cytometry is based primarily on analyte size and light scattering characteristics, and the use of fluorescent probes. Fluorophores based cellular chemistry or fluorophore-conjugated antibodies may be used, including FITC, PE, APC, PerCP, Cy3/5, and Alexa, with different absorbance and emission spectra. The use of these fluorophores allows us to analyze multiple parameters in a single assay (Brown and Wittwer, 2000). The process basically involves the aspiration of a suspension of cells or particles into a narrow stream allowed to pass one at a time through a focused laser beam, which may be either scattered or absorbed. The absorbed light is re-emitted as fluorescence after it detects the fluorochrome labeled antibody that links with the surface of the cell or internal cell structure. These fluorescence signals are detected by a series of photodiodes, amplified as electrical pulses which are then digitized. The digitized data for each and every cell are analyzed quantitatively and statistically. Besides real-time characterization of several thousands of particles, a flow cytometer can also segregate and isolate particles based on specified properties, a technique better known as fluorescence-activated cell sorting (FACS) using specific fluorescent characteristics of each cell (Brown and Wittwer, 2000).

## 3.3 CLINICAL APPLICATIONS OF FLOW CYTOMETRY

Recent years have seen a rise in the use of flow cytometry for ***fetal red cells detection and quantification in maternal blood***. The mother's blood is tested for the presence and quantity of fetal red cells in case of suspected fetomaternal hemorrhage, and an appropriate amount of Rh-immune globulin is administered. Although there are quantitative tests like the Kleihauer-Betke acid-elution test, it shows interlaboratory variability, is time-consuming and labor-intensive (Polesky and Sebring, 1981). Compared to the Kleihauer-Betke test, For the studies on fetal cells, flow cytometry serves as a more reproducible, and sensitive technique. The method here can either use fluorescently labeled antibodies to the rhesus (D) antigen, or hemoglobin F antibodies (Brown and Wittwer, 2000; Davis et al. 1998) and can distinguish fetal cells from the adult red cells with small amounts of hemoglobin F.

The ***peripheral blood reticulocytes counting*** is important in observing bone marrow regenerative activity after chemotherapy or bone marrow transplantation (Viana et al, 2014). The technique relies on the presence of RNA and other substances in the reticulocyte, which are lost while differentiating into mature red blood cells. Flow cytometry has a definite advantage over manual counting in being is more accurate, precise, and cost-effective, and proves to be valuable in the diagnosis and monitoring of diseases (Garratty G, Arndt, 1999).

An acquired clonal stem cell disorder, ***paroxysmal nocturnal hemoglobinuria (PNH)*** leads to intravascular hemolysis accompanied with complications related to thrombosis and infections. Aplastic anemia may give rise to PNH, which may be followed by acute leukemia (Peters and Ansari, 2011). Conventionally, PNH can be diagnosed by sugar water test and the Ham's acid hemolysis test (Preis and Lowrey, 2014) which show low specificity and have stringent specimen criteria. The decay-accelerating factor and membrane-inhibitor of reactive lysis have specific antibodies CD55 and CD59, respectively, and can be used for a definitive diagnosis of PNH by flow cytometry (Dunphy, 1999). The affected patients display different degrees of expression of CD55 and CD59, leading to the easy identification of two or more populations of erythrocytes. For the flow cytometric analysis and counting of reticulocytes, fluorescent dyes such as thiazole orange are used that bind the residual RNA. The method can thus very precisely differentiate between reticulocytes and mature RBCs that are mature, with better sensitivity, and reproducibility compared to the classical method. But a shortcoming is that the remnant of nuclear DNA, also called the Howell-Jolley bodies are not distinguished from reticulocytes (Brown and Wittwer, 2000).

Flow cytometry can be used as a complementary or replacement for immunological testing of red cells. The ***recipient's red cell phenotype can be accurately identified*** by flow cytometry (Garratty G, Arndt, 1999). It is very difficult to determine the blood type of the recipient with multiple transfusions. In blood banks, leukocyte contamination in leukocyte-reduced blood products are assessed by flow cytometry (Sharma and Marwaha, 2010).

In ***the case of HIV***, the backbone of a clinical flow cytometry laboratory is leukocyte analysis and Immunologic monitoring of HIV-infected patients. The helper/inducer T lymphocytes are infected by HIV via the CD4 antigen. As HIV disease progresses, there is a decrease in total CD4-positive T lymphocytes. Thus, the absolute CD4 count is an important measurement for predicting, and monitoring the progression of disease and treatment response in HIV-infected individuals (Obirikorang and Yeboah, 2009). Therefore, CD4 counts assess the patient's immune status, besides determining the amount of virus; both pieces of informations may be used for diagnosis, prognosis, and therapeutic management of the disease.

***The distinction between lymphoid and myeloid lineages*** in acute and minimally differentiated leukemias can be effectively done by flow cytometry. While it is difficult to classify most acute myeloid leukemias only on the basis of phenotype, flow cytometry still enables distinction of acute myeloid leukemias, such as acute promyelocytic leukemia (Dong et al. 2011). Therapy-resistant leukemias can also be identified by flow cytometry, and the outcome has been shown to correlate strongly in ALL (Chiaretti et al. 2014). Often, specific antigen patterns are observed in B-cell lymphoproliferative disorders, which allow the clinicians to make the use of a wide range of antibodies and make specific diagnoses.

***Clonality in B-cell populations*** can be effectively identified by flow cytometry. Acting as a powerful adjunct to cytologic diagnosis, the lymphoma phenotyping of fine needle aspirates can be successfully done by flow cytometry (Dong et al. 2011). Because the technique is sensitive and can simultaneously analyze multiple characteristics, flow cytometry becomes useful for the detection of minimal residual disease, particularly in cases of abnormal patterns of antigen expression (Peters and Ansari, 2011).

In both research and clinical laboratories, ***platelet analysis*** by flow cytometry is becoming increasingly common. Platelet-associated immunoglobulin assays by flow cytometry can be assayed directly or indirectly, just like other platelet-associated immunoglobulin immunoassays (Huh et al. 2009).

The malignant cells may persist at levels below the conventional detection limit and may be a source of disease relapse in many patients. Flow cytometry acts as a simple and rapid method for detection of ***minimal residual disease (MRD)***, and the persistence of malignant cells even at levels much lower than the conventional detection limits (Choudhury et al. 2013). The criteria for detection of MRD include sensitivity, specificity, reproducibility, and easy standardization and rapidity (Theunissen et al. 2017). While flow cytometric analysis is less sensitive than the polymerase chain technique for MRD, it is simple and rapid to perform, provides quantitative data, and has adequate sensitivity. In the sample under study, flow cytometry detects immunophenotypic features that are aberrant and are not exhibited by normal cell populations.

The progress of patients after chemotherapy or transplantation can be sensitively and effectively monitored by flow cytometry, which can also detect the minimal residual disease (MRD). According to the World Health Organization (WHO) guideline, current practice in the diagnosis of leukemia is based on incorporating data obtained from flow cytometry analysis, in addition to molecular as well as karyotypic data.

Currently, flow cytometry is used to ***determine the clonal proliferation*** (B or T cell), for the diagnosis of hematological and lymphoid neoplasia. However, flow cytometry is a support to the microscopic examination of cells and the clinical history.

Even though flow cytometric analysis is less sensitive (one neoplastic cell for every 103–104 normal cells) for MRD than the quantitative polymerase chain reaction (qPCR) technique, it is simple and rapid to perform, provides quantitative data, and has adequate sensitivity in many leukemia cases. Several clinical correlative studies have been able to predict the outcome by flow cytometric detection method using MRD leukemia-specific markers (Dong et al. 2011; Sharma and Marwaha, 2010). In all these studies, flow cytometric panels of antibodies were used to define unique immunophenotypic signatures of leukemic cells that distinguish leukemic blasts from their normal counterparts. This concept was clearly demonstrated when leukaemic and normal cells were analysed resulting in discerning normal (CD19+, CD10+, CD34+) and leukaemic phenotypes (CD38+, CD97+, CD86+), allowing detection of ALL cells at a level of 0.001% (detection of 1 leukaemic cell among 105 bone marrow cells) (Choudhury et al. 2013).

Thus, multi-parameter flow cytometry has a generally high predictive value for relapse, which makes it a logical and promising tool for assessment of MRD.

## 3.4 DETERMINING THE EFFECTIVENESS OF A FEW THERAPIES

A major problem faced by oncologists is the selection of an optimal chemotherapeutic agent. Individual variability in tumors often leads to an inefficient therapeutic result.

The patients may develop ***multiple drug resistance (MDR),*** and cause chemotherapy to fail. While over-expression of P-gp, an MDR1 gene product, and other proteins involved in cellular transport frequently cause MDR, there appear to be other mechanisms like detoxification by biochemical means, as well as DNA replication and repair (Theunissen et al. 2017). In conjunction with immunocytochemistry and molecular techniques, flow cytometry has been an essential tool for determining the expression of markers of MDR, as well as assessing the effectiveness of chemotherapeutic drugs (Pisco et al. 2014).

Flow cytometry has multi-parameter capabilities, and is thus ideally ***suited to assess the apoptotic regulatory status***. A study encompassing patients presenting with late-stage (IIIB and IV) small cell lung cancer and non-small cell lung cancer identified the subset of patients who had a relatively higher likelihood of an apoptotic response to platinum-based chemotherapy (Ozkaya et al. 2012). In a very recent study, Bradford and Khan (2013) investigated the efficacy of metformin as a chemosensitizer, using flow cytometric analysis and studied a total of 14 solid tumors of various types. Metformin along with chemotherapeutic agent had a synergistic effect, resulting in a high sensitivity to apoptosis.

Flow cytometry can also be used to detect the ***sensitivity of tumor cells to radiation*** prior to therapy, to determine apoptosis. This method has been used to design protocols for improved, accurate and reliable drug treatment.

***Determination of cell viability or apoptosis*** using flow cytometry aids in drug designing and treatment protocols to improve the accuracy and reliability of the conventional assay. One of the most rapidly growing areas of therapeutics is molecule-targeted, or ligand or antigens using monoclonal antibodies. While some of these agents may directly disrupt cell proliferation, may be anti-apoptotic, others may target other cytotoxic products (Patra et al. 2016). It is important to do flow cytometric analysis before the treatment to ensure that the aberrant cells express the specific antigen, or to verify binding of the antibody and to determine the tumor cell eradication efficiency during and after treatment. The binding of antibody and removal of tumor cells can be monitored by flow cytometry, both during and after the treatment (Bannas et al. 2015).

***Organ transplant recipients*** are known to frequently suffer from toxicity, or the immunosuppressive drugs may not be efficient. These responses can be attributed to individual variations in drug sensitivity. This variable component can be resolved through the use of pharmacodynamic monitoring that focuses on measuring the biological effects of drugs. In this connection, one may mention phospho-specific flow cytometry, which can monitor not only post-translation levels, intracellular cytokine production, and cytotoxic and proliferative activities (Carla et al., 2012), but can also determine the activity of intracellular immune signaling. In fact, Dieterlen et al. (2012) were amongst the first to show the potential of pharmacodynamic monitoring of mTOR inhibitor molecules by flow cytometry.

***Lymphocyte subset enumeration in AIDS patients*** Human immunodeficiency virus (HIV) infection is characterized by the depletion of CD4 T cells, expansion of CD8 T cells and chronic activation of the immune system, which lead to immune dysfunction. The response to the depletion of CD4 T cells and the inflammatory response to HIV infection are reflected in CD4 counts and in HIV RNA levels, respectively. Flow cytometry is an accepted standard method for the determination of the absolute count of CD4+, CD3+ and CD8+ T lymphocytes (Obirikorang C, Yeboah, 2009), and has proven critical for the diagnosis of HIV infection, as well as the management of patients receiving antiviral therapy.

Stem cells have the potential to revolutionize tissue regeneration and engineering. They can be useful in understanding general cellular processes, e.g. embryogenesis, organogenesis, cancer, and aging. Further, they have been used as a vehicle for the generation of transgenic mice for the purposes of functional gene analysis and the construction of disease models. Stem cells are also used in industrial research for cell-based screening in the discovery of drug targets or drugs, or for predictive toxicology, and by clinical researchers looking for a potential source for tissue regeneration. A review by Preffer and Dombkowski (2009) outlined the important surface and functional antigenic markers of stem cells, describing their progenitor expressions and respective differentiated cell expression (Preffer and Dombkowski, 2009).

Hordyjewska et al. (2015) reported the use of fluorochrome-conjugated monoclonal antibodies directed against the CD34 molecule to identify CD34+ cells by flow cytometry to study the status of hematopoeitic cells.

Similarly, Hoechst 33342 dye, known to bind to the A-T rich regions of DNA, is used to detect and isolate stem and early progenitor cells, that is identified as a side population (SP) cells by their low Hoechst blue and red fluorescent emission signature (Preffer and Dombkowski, 2009).

## 3.5 FLOW CYTOMETRY: SCOPE OF ITS APPLICATIONS

The ability to sort cells with specific features opens the door to molecular analysis of highly defined populations. Whether it is in the area of analyzing the complex surface markers of a lymphoid malignancy or the cytoplasmic pattern of apoptosis regulation (as mentioned in earlier sections), flow cytometry has become far more complex than a simple fluorescence identifying instrument. For example, the quantification of mitochondrial membrane potential or the activity of intracellular enzymes by methods such as ELISA measures secreted proteins produced by entire cell populations. In contrast, intracellular flow cytometry allows for not only the analysis of cytokines and other inflammatory mediators produced by individual cell types, but also makes it possible to easily determine if the cytokine production by an activated cell population is the result of a few cells producing large amounts of cytokine or a large population of cells producing small quantities of cytokine per cell (Smith et al. 2015).

Another example of how the applications of flow cytometry can be broadened is the use of protein transport inhibitors (commonly, monensin and brefeldin A) to ***detect cytokines trapped within the cell*** (Smith et al. 2015).

Yet another area in which flow cytometry can be used is ***the analysis of cell function,*** which can provide relevant clinical information, such as on the expression of surface antigens. Through flow cytometry, it is possible to measure nearly every event the lymphocytes are being activated. Some processes include the determination of calcium flux, tyrosine phosphorylation, neoantigen expression, oxidative metabolism, and cellular proliferation. Schepers et al. (2009) described a flow cytometric method allowing for the simultaneous kinetic analysis of changes in cytoplasmic calcium in monocytes, granulocytes, and lymphocytes (Schepers et al, 2009).

After a decade of the practical application of flow cytometry, there now exists a large number of reports on human studies in the area of pure basic research as well as in medical applications in clinics. These reports demonstrate that flow cytometry provides valuable clinical information when combined with other diagnostic procedures. Presently, several approaches appear to be rather promising. However, the malleability of the technique has resulted in the use of a vast array of approaches that has led to problems in data reproducibility. It is important to explore new applications of flow methods using evolving technologies to improve the diagnostic and prognostic value of flow cytometry.

Flow cytometry along with information about human cell cycle has enabled the development of ***DNA analysis of neoplasia*** (O'Donnell et al. 2013). For this, a specific number of cells are stained with a fluorescent dye which is DNA-specific under controlled conditions. The cells are then flow cytometrically analyzed in a period of a few minutes, the fluorescent amount and intensity of the dye bound to each cellular DNA is measured based on a statistically significant number of cells. Moreover, in a heterogeneous mixture, small specific cell populations can be detected.

The ***analysis of platelet structure and function*** has been done in several clinical assays by specialized flow cytometry laboratories and are poised to be practiced more in the near future as standardized techniques and controls become available (Huh et al. 2009).

***Cell Function Analysis*** can provide relevant clinical information like the expression of surface antigens which cannot be acquired from static cellular parameters. The information obtained are especially relevant in transplantation medicine and immune diseases, thus making lymphocyte analysis relevant. While flow cytometry can measure nearly every event that occurs during the process of lymphocyte activation, the status which has the greatest clinical potential are oxidative metabolism, tyrosine phosphorylation, calcium flux, cellular proliferation, and neoantigen expression (Brown and Wittwer, 2000).

Flow cytometry aids in ***tyrosine phosphorylation measurement*** within the cell by using monoclonal antibodies specific for different tyrosine kinase substrates. These monoclonal antibodies may be either directly labeled or may be linked to colored microbeads (Xie et al.2009).

In the case of chronic granulomatous disease, ***measurement of the oxidative burst in neutrophils*** by flow cytometry has been used as a reliable screening test. For this purpose, a dye that does not fluoresce, for example, dihydrorhodamine-123 is commonly used, which, during normal oxidative burst oxidizes to a brightly fluorescent compound rhodamine-123 (Brown and Wittwer, 2000).

## 3.6 APPLICATIONS IN MICROBIOLOGY

Flow cytometry has been shown to be valuable in clinical microbiological applications. Classical microbiology techniques are relatively slow, mainly due to the need to culture the microorganisms. Moreover, it becomes especially difficult to analyze the microorganisms that are difficult to culture. Interestingly, during World War II an aerosol particle counter was used by the US Army. This counter is an ancestor of the modern flow cytometer to analyze mine dust, bacteria, and spores, in attempts to detect biowarfare agents (Álvarez-Barrientos et al. 2000). Modern flow cytometry uses specific cytometric parameters or certain fluorochromes and allows single- or multiple- microbe detection (bacteria, fungi, parasites, and viruses) in an easy and reliable manner.

It must be noted that fluorescent staining is the hallmark of flow ***cytometric discrimination of microorganisms.*** The staining substances include dyes that are sensitive to membrane potential, nucleic acid stains, and fluorogenic enzyme substrates. Flow cytometry also enables development of rapid, accurate, and highly reproducible procedures to assess drug susceptibility and cytotoxicity, as well as the presence of heterogeneous populations.

To ***discriminate among different bacteria in the same sample***, Steen et al applied dual-parameter analysis, which included scattered light, and the other was either autofluorescence or fluorescence emission from fluorochromes coupled to cellular components like protein and DNA, or light scattered acquired from another angle. Despite these developments, the simplest way to visualize or identify microorganisms by flow cytometry involve either direct fluorochrome staining or through antibody or oligonucleotide conjugates (Ambriz-Aviña et al. 2014).

Flow cytometry enables ***assessment of microbial viability*** simply and rapidly. Several groups have studied the bacterial membrane potential under environmental stress or starvation by using fluorochromes that differentiate nonviable, viable, and dormant cells (de Jong et al. 2012).

***Studying microbial metabolism*** is another aspect of flow cytometry. The first attempt towards this was made by using NADPH and flavins mediated autofluorescence markers. Other studies have based on intracellular pH, determination of DNA and proteins, peroxide production, and (Ambriz-Aviña et al. 2014).

Flow cytometry-based counting of live and dead gram-positive and gram-negative bacteria and yeasts has been made possible by recently developed kits and fluorochromes like Sytox Green and Live/Dead kits (Molecular Probes, Eugene, Oreg.) (Ghosh et al. 2013), which significantly simplify staining procedures and making data interpretation easier. The Gram stain reaction of clinically relevant strains can be successfully predicted and classified. These kinds of studies are facilitated more with the development of fluorescent protein-based gene expression systems (Ambriz-Aviña et al. 2014).

Using this technique, Philips and Martin were able to successfully detect *Bacillus* spores (Álvarez-Barrientos, 2000), and spores from *Dictyostelium discoideum* and *Enterocytozoon bieneusi* could be sorted by Griffiths et al. (Álvarez-Barrientos, 2000) and Challier et al. (Álvarez-Barrientos, 2000), respectively.

## *Direct Detection of microbes*

The surface antigens of *Mycobacterium*, *Haemophilus*, *Salmonella*, *Mycoplasma*, *Brucella*, *Pseudomonas*, *Legionella*, and *Bacteroides* have been successfully detected by flow cytometry using fluorescent antibodies (Álvarez-Barrientos et al. 2000). In heterogeneous populations, the use of antibodies ensures sensitivity and specificity in the detection of cell types as few as 100 cells per mL in 30 min (Clarke et al. 1998).

Fluorescent-oligonucleotide detection is yet another flow cytometry technique that can enable direct diagnosis. Wallner et al. (1997) combined fluorescent probes that are rRNA-targeted and 4′,6-diamidino-2-phenylindole (DAPI) for nucleic acid staining, and successfully detected *Acinetobacter* species. However, while the method is complex and is not used for testing clinical samples, it provides significant specificity to identify a putative infectious agent (Álvarez-Barrientos, 2000), and still holds promise for future applications.

### *Detection of Fungi*

The treatment outcomes based on conventional techniques like clinical presentation, culture, and microscopy-based diagnosis may be hindered by false-negative and false-positive results. Mercure et al. (1996) validated a reliable serotyping procedure by flow cytometry, using serotype A-specific antisera. They analyzed the isolated strains by flow cytometry and were able to ascertain the presence of two different strains in the culture. In another report, Pierard et al. (1996) were able to identify and distinguish pathogenic fungus from nonpathogenic ones by flow cytometry. His group could establish the treatment in cases of mixed infections (Álvarez-Barrientos et al. 2000).

### *Identification of Parasites*

Several approaches based on the absence of DNA in erythrocytes have been developed in the past to detect intracellular parasites (Álvarez-Barrientos et al. 2000). Thus, it is possible to stain and detect the parasite DNA with specific fluorochromes and if the parasite is inside the cell. Flow cytometry also enables to other characteristics, such as parasite antigens expressed by the erythrocyte (Jouin et al. 1995) or state of viability of the host cell. Flow cytometry can be used along with immunofluorescence, and analyze a larger number of samples in a comparatively short period. Importantly, when samples containing a small number of cysts were analyzed, the technique was more consistent than conventional tests or immunofluorescence microscopy.

### *Detection and quantification of viruses*

Flow cytometric detection of viral antigens is a useful assay for the clinical diagnosis of several blood-borne viruses. Flow cytometry can detect and quantify virus-infected cells using specific antibodies against the surface or internal antigens (Álvarez-Barrientos et al. 2000; McSharry et al. 1990). In the case of internal antigens, the cells need to be permeabilized. The method may use a direct method, which uses fluorescently labeled antibody or indirect fluorescent-antibody methods.

Analysis of viral markers is important for the antiviral treatments (Álvarez-Barrientos et al. 2000). Flow cytometry has been successfully used for the detection of viral antigens for diagnosing infections due to hepatitis and herpesvirus. The technique involves using antibodies for the detection of HBV surface (HBsAg) and HBV core (HbcAg) antigens in peripheral blood mononuclear cells from HBV infected patients (Álvarez-Barrientos et al. 2000; Roisman et al. 1994). Flow cytometry is also able to detect Hepatitis C virus (HCV), woodchuck hepatitis virus, and varicella-zoster virus in peripheral blood mononuclear cells using monoclonal antibodies to HCV core antigen, polyclonal antibodies to woodchuck hepatitis virus, and antibodies to the gpI glycoprotein from the varicella-zoster virus (Álvarez-Barrientos et al. 2000).

### *Viral nucleic acid detection and quantification*

In the case of active CMV infection, identification of specific CMV-DNA or RNA by flow cytometry may permit possible phenotyping of cells. Flow cytometry can detect and quantify mononuclear peripheral blood leukocytes in CMV infected patients (Álvarez-Barrientos et al. 2000; Link et al, 1992), and in differentiating cells that harbor actively replicating virus and the cells with latent infection.

## 3.7 SUMMARY

The flow cytometer has unique analytic capabilities and is a versatile tool with vast potential for cells and particle studies, and has become an integral part of the medical research in the past decade. While flow cytometric studies are performed widely in clinical laboratories, the other specialized applications are limited to larger labs because of economic constraints. While flow cytometry has several advantages, it is not widely used in clinical microbiological studies. This may be due to limited access to flow cytometers or due to inadequate information about the potential of this technique. Nevertheless, the flow cytometer or its variant techniques are valuable for medical diagnosis and the following therapy. New fluorochromes, and complex of dyes are under development, in addition to advanced software.

## 3.8 REFERENCES

1. Álvarez-Barrientos A, Arroyo J, Cantón R, Nombela C, Sánchez-Pérez M. Applications of Flow Cytometry to Clinical Microbiology. Clinical Microbiol Rev. 2000;13(2):167-195.
2. Ambriz-Aviña V, Contreras-Garduño JA, Pedraza-Reyes M. Applications of Flow Cytometry to Characterize Bacterial Physiological Responses. BioMed Research Int. 2014; 2014, Article ID 461941, 14 pages.
3. Bannas P, Lenz A, Kunick V, Fumey W, Rissiek B, Schmid J, et al. Validation of Nanobody and Antibody Based *In Vivo* **Tumor Xenograft NIRF-imaging Experiments in Mice Using** *Ex Vivo* **Flow Cytometry and Microscopy.** J Vis Exp. 2015; (98): 52462.
4. Bradford SA, Khan A**.** Individualizing Chemotherapy using the Anti-Diabetic Drug, Metformin, as an "Adjuvant": An Exploratory Study. J Cancer Sci Ther. 2013; 5:120-125. doi: 10.4172/1948-5956.1000197
5. Brown M, Wittwer C. Flow Cytometry: Principles and Clinical Applications in Hematology. Clinical Chem. 2000; 46:8(B) 1221–1229.
6. Carla B, Anne B, Ramin V, Willem W. Phospho-specific flow cytometry for pharmacodynamic monitoring of immunosuppressive therapy in transplantation. Transplant Res. 2012; 1:20. doi: 10.1186/2047-1440-1-20.
7. Chiaretti S, Zini G, Bassan R. Diagnosis and Subclassification of Acute Lymphoblastic Leukemia. Mediterr J Hematol Infect Dis. 2014; 6(1): e2014073.
8. Choudhury Sangeeta, Dheeraj M, Veronique D, Vibha T, Shishir A, Keerti M, Kriti J, Nirmal KG. (2013). Applications of flow cytometry in healthcare: An update. The Ganga Ram Journal. 2013; 3:158–169.
9. Clarke RG, Pinder AC. Improved detection of bacteria by flow cytometry using a combination of antibody and viability markers. J Appl Microbiol. 1998; 84:577–584.
10. Davis BH, Olsen S, Bigelow NC, Chen JC. Detection of fetal red cells in fetomaternal hemorrhage using a fetal hemoglobin monoclonal antibody by flow cytometry. Transfusion. 1998; 38:749-756.
11. de Jong IG, Veening J, Kuipers OP. Single cell analysis of gene expression patterns during carbon starvation in Bacillus subtilis

reveals large phenotypic variation. Environ Microbiol. 2012; 14(12), 3110–3121.

12. Dieterlen MT, Bittner HB, Klein S, von Salisch S, Mittag A, Tárnok A, Dhein S,Mohr FW, Barten MJ. Assay validation of phosphorylated S6 ribosomal protein for a pharmacodynamic monitoring of mTOR-inhibitors in peripheral human blood. Cytometry B Clin Cytom. 2012, 82:151–157.
13. Dong HY, Kung JX, Bhardwaj V, McGill J. Flow Cytometry Rapidly Identifies All Acute Promyelocytic Leukemias With High Specificity Independent of Underlying Cytogenetic Abnormalities. *Am J Clin Pathol. 2011;* 135(1), 76–84.
14. Dunphy CH. Comprehensive review of adult acute myelogenous leukemia: cytomorphological, enzyme cytochemical, flow cytometric immunophenotypic, and cytogenetic findings. J Clin Lab Anal 1999;13(1):19-26.
15. Garratty G, Arndt PA. Applications of flow cytofluorometry to red blood cell immunology [In Process Citation]. Cytometry. 1999; 38:259-267.
16. Ghosh S, Indukuri K, Bondalapati S, Saikia AK, Rangan L. Unveiling the mode of action of antibacterial labdane diterpenes from Alpinia nigra (Gaertn.) B. L. Burtt seeds. Eur J Med Chem. 2013; 66, 101–105.
17. Hordyjewska A, Popiołek L, Horecka A. Characteristics of hematopoietic stem cells of umbilical cord blood. Cytotechnology. 2015 May; 67(3): 387–396.
18. Huh HJ, Park CJ, Kim SW, Han SH, Jang S, Chi HS. Flow Cytometric Detection of Platelet-Associated Immunoglobulin in Patients with Immune Thrombocytopenic Purpura and Nonimmune Thrombocytopenia. Ann Clin Lab Sci. 2009; 39(3).
19. Jouin H, Goguet de la Salmoniere YO, Behr C, Huyin QD, Michel JC, Sarthou JL, Pereira DS, Dubois P. Flow cytometry detection of surface antigens on fresh, unfixed red blood cells infected by *Plasmodium falciparum*. J Immunol Methods. 1995; 179:1–12.
20. Link H, Battmer K, Kleine HD. Detection of cytomegalovirus-infected cells by flow cytometry and fluorescence in suspension hybridisation (FLASH) using DNA probes labeled with biotin by polymerase chain reaction. J Med Virol. 1992; 37:143–148.
21. McSharry JJ, Costantino R, Robbiano E, Echols R, Stevens R, Lehman JM. Detection and quantitation of human immunodeficiency virus-

infected peripheral blood mononuclear cells by flow cytometry. J Clin Microbiol. 1990; 28:724–733.

22. Mercure S, Senechal S, Auger P, Lemay G, Montplaisir S. *Candida albicans* serotype analysis by flow cytometry. J Clin Microbiol. 1996; 34:2106–2112.
23. Obirikorang C, Yeboah FA. Blood haemoglobin measurement as a predictive indicator for the progression of HIV/AIDS in resource-limited setting. J Biomed Sci. 2009; 16(1): 102.
24. O'Donnell AE, Ernst DN, Hingorani R. Multiparameter Flow Cytometry: Advances in High Resolution Analysis. Immune Netw. 2013 Apr; 13(2): 43–54.
25. Ozkaya S, Findik s, Dirican A, Atici AG. Long-term survival rates of patients with stage IIIB and IV non-small cell lung cancer treated with cisplatin plus vinorelbine or gemcitabine. Exp Ther Med. 2012 Dec; 4(6): 1035–1038.
26. Patra B, Peng C-C, Liao W-H, Lee C-H, Tung Y-C. Drug testing and flow cytometry analysis on a large number of uniform sized tumor spheroids using a microfluidic device. Sci Rep. 2016; 6: 21061.
27. Pierard GE, Arrese JE, De Donker P, Pierard-Franchimont C. Present and potential diagnostic techniques in onychomycosis. J Am Acad Dermatol. 1996; 34:273–277.
28. Pisco AO, Jackson DA, Huang S. Reduced intracellular drug accumulation in drug-resistant leukemia cells is not only solely due to MDR-mediated efflux but also to decreased uptake. Front Oncol. Oct 2014.
29. Polesky HF, Sebring ES. Evaluation of methods for detection and quantitation of fetal cells and their effect on RhIgG usage. Am J Clin Pathol. 1981; 76:525-529.
30. Preffer F and Dombkowski D. Advances in Complex Multiparameter Flow Cytometry Technology: Applications in Stem Cell Research. Cytometry B Clin Cytom. 2009 Sep; 76(5): 295–314.
31. Roisman FR, Castello A, Fainboim H, Morelli A, Fainboim L. Hepatitis B virus antigens in peripheral blood mononuclear cells during the course of viral infection. Clin Immunol Immunopathol. 1994; 70:99–103.
32. Schepers E, Glorieux G, Dhondt A, Vanholder R. Flow cytometric calcium flux assay: Evaluation of cytoplasmic calcium kinetics in

whole blood leukocytes. J Imm Methods. 2009; 348(1-2):74-82·

33. Sharma RR, Marwaha N. Leukoreduced blood components: Advantages and strategies for its implementation in developing countries. Asian J Transfus Sci. 2010; Jan; 4(1): 3–8.

34. Smith SG, Smits K, Joosten SA, van Meijgaarden KE, Satti I, Fletcher HA, et al. Intracellular Cytokine Staining and Flow Cytometry: Considerations for Application in Clinical Trials of Novel Tuberculosis Vaccines. PLoS ONE. 2015; 10(9): e0138042. https://doi.org/10.1371/journal.pone.0138042

35. Theunissen P, Mejstrikova E, Sedek L, Sluijs-Gelling AJ, Gaipa G, Bartels M et al. Standardized flow cytometry for highly sensitive MRD measurements in B-cell acute lymphoblastic leukemia. Blood. 2017 Jan 19; 129(3): 347–357.

36. Viana KA, Maria das GC, Sant'Ana Dusse LM, Fernandes AC, Avelar RS, Avelar DMV, et al. Flow cytometry reticulocyte counting using acridine orange: validation of a new protocol. J Bras Patol Med Lab. 2014; 50(3).

37. Wallner G, Fuchs B, Spring S, Beisker W, Amann R. Flow sorting of microorganisms for molecular analysis. Appl Environ Microbiol. 1997; 63:4223–4231.

38. Xie S, Moya C, Bilgin B, Jayaraman A, Walton SP. Emerging Affinity-Based Techniques in Proteomics. Expert Rev Proteomics. 2009 Oct; 6(5): 573–583.

# CHAPTER 4

# FLOW CYTOMETRY IN A CLINICAL SCENARIO

**Shiv Sanjeevi Sripathi**

Shiv Sanjeevi Sripathi secured second rank in Mumbai University and first rank in his college at his undergrad level and in his college during his Masters and has worked on a stem cell project at the Specialized Centre for Cell Based Therapy (SCCT), KEM Hospital at Mumbai, following which he was a Junior Research Fellow at Microbiology & Cell Biology Department, Indian Institute of Science, Bangalore from August 2009 to November 2011. The field of work involved cloning of cell wall genes and transcription factors in E. coli & M. smegmatis. This was followed by a stint as a Research and Development Executive at Transasia Biomedicals, Mumbai where he worked on patent writing and Immunoassays from 2011 till 2013. In 2013, he entered the teaching field where he has been tutoring biology and chemistry to the students of 11th and 12th grades along with English grammar and communication skills to people of all age groups. Shiv also loves helping people write scientific manuscripts and has co-edited several books on life sciences and aspects of biotechnology with Apple Academic Press.

## 4.1 INTRODUCTION

The technique of flow cytometry had its origins in 1960 while it was accepted in a clinical scenario in the early stages of 1980 where its application was for managing HIV patients. The instrument was used in diagnosis of the disease as the pathogen behind the disease had not been identified at that time. Research now uses this approach to look at such diseases of the immune system for possible mechanisms. Flow cytometry has ushered in a change in the scene in the case of pathology and hematology. The technique has become part of diagnosis of diseases such as leukaemia as well as its role as a routine approach in a modern day lab (Virgo and Gibbs, 2012).

## 4.2 FLOW CYTOMETRY IN MICROBIOLOGY

One of the major challenges with reports from microbiological techniques is the time factor. The culture of bacteria or fungi or virus requires time before their analysis. While the techniques for such analysis have seen advancements to make them sensitive the time is an important factor as several diseases need quick and prompt treatments. Most labs face competition that requires quick results that can be achieved by automation. Such an example can be seen by systems that screen for urinary tract infections or susceptibility testing of anti-microbials. However, constant modifications are the need of the hour that can enable increased sensitivity and faster results.

Such a need can be achieved with the use of flow cytometry. Such a system is pitched at serving several purposes:

- Detect organisms that cause a pathological condition
- Identification of the organism
- Several slow growing organisms (fungi/ Mycobacteria) are identified faster: for instance a span of hours can show the results.
- In mixed populations, the effects of different microbes are different; making quick analysis possible.

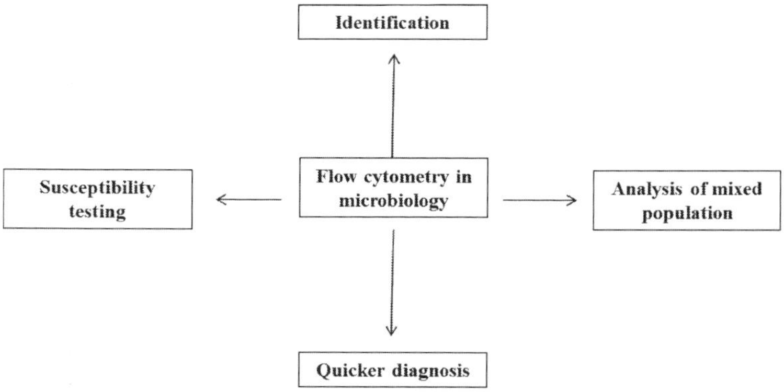


**Figure 4.1:** Flow cytometry and clinical microbiology.

- detection of antibodies in patients
- Monitoring of clinical status post therapy
- The application of specific parameters can aid in aid in circumventing subjectivity to facilitate interpretation of results (Alvarez Barrientos et al, 2000).

The association of flow cytometry and microbiology has dated back a long time in history. An early construct of a flow cytometer was called aerosol particle counter that was employed for enumeration of microbes in the Second World war by the US. We can see that the use of flow cytometric system for microbes began earlier than that of its use for current studies such as mammal samples.

The study of properties can be aided with the use of fluorochromes: there are several classes of these molecules that include those that bind molecules such as Nile red that binds lipids or propidium iodide that binds nucleic acids and fluorescein isothiocyanate that reacts with proteins as well as fluorescence detected using various properties of enzymes and cells. Let us see a few of the molecules in table 4.1.

**Table 4.1:** Dyes for slow cytometry of microbes

| Dye | Excitation and emission wave-lengths (nm) | Binding molecule | Use |
|---|---|---|---|
| SYTOX Green | 504/525 | DNA/RNA | Viability<br>Quantification of nucleic acids |
| Ethidium bromide | 510/595 | DNA/RNA | Study of cell cycle<br>Quantification |

| | | | |
|---|---|---|---|
| Propidium iodide(PI) | 536/625 | DNA | Viability<br>Quantification of nucleic<br>Study of cell cycle |
| Hoechst | 340/450 | GC rich areas of DNA | Study of cell cycle |
| Pyronine Y | 497/563 | RNA | Quantification |
| Texas Red (sulforhodamine isothiocyanate) | 580/620 | Protein | Detection |
| FITC | 495/525 | | Detection |
| Lectins | - | Sugars | Composition of cell wall<br>Detection |
| SNARF-1 | 510 /587–635 | pH detection | Metabolism |
| Antibodies | Flourochrome (any) | Antigen/ Specific proteins | Detection |
| Oligonucleotides | Flourochrome (any) | Specific sequence | Detection |

## 4.3 MILESTONES FOR FLOW CYTOMETRY IN A CLINICAL SETTING

While the technique originated as an analysis of single cells flowing in a fluid using laser illumination, it has seen several advancements in fluorescent dyes (described above) or the use of antibodies that target specific antigens.

The year 1972 saw Herzenberg and colleagues invent FACS (Flourescent Activated Cell Sorting) followed by chromosomes sorted by flow cytometry in the Human Genome Project. Let us see how the tool has aided diagnosis of several diseases (Table 4,2) (Jaye et al, 2012).

**Table 4.2:** Flow cytometry and disease diagnosis

| Year | Disease | Use |
|---|---|---|
| 1980s | Cancer | Diagnosis of lymphomas or leukemia<br>Prognosis studies |
| 1980s | AIDS | Looking at CD4 populations |
| 1980s | Immune deficiencies | Measuring B and T cells |
| 1990s | Cancer | Looking for remainder cancer cells post treatment |
| 1990s | Cell based therapy | Counting CD34 or stem cells to look at efficiency of a bone marrow transplantation |
| 2000s | Transplantation | Using antibodies to HLA in patients to look at graft-versus-host disease |

## 4.4 CASE STUDY: FLOW CYTOMETRY AND DIAGNOSIS

The characters of cells along with their functions and properties can be assessed with the use of flow cytometry as a key feature of a cell called cluster of differentiation (CD) is used. Let us see an example of such an application of flow cytometry in diagnosis:

A patient with a history of breast cancer that was diagnosed in 2006 followed by elevated BP in 2010 following chemotherapy and radiation for her cancer; there was no other complaint or family history. However this 63 year lady reported loss of appetite, weight loss, bruising and fatigue. The patient use to take rest thrice a day that however made no difference to her tiredness; while there she also reported of continuous joint pain. Following examination the reports were:

- 38.9°C
- 108 heart beats per minute
- 158/68 BP and pale oral mucosa
- Blood tests showed 110,000 x $10^3$/µL leucocyte count

Flow cytometry of bone marrow samples showed 22.6% $CD34^+/CD117^+$. This aided the diagnosis of acute myeloid leukemia (AML).

In the case of acute lymphoblastic leukemias (ALLs) and AMLs the cells have several subtypes that differ in structure and phenotype. For example, as seen in the above example: the markers for AML precursors are CD34, CD38, CD117! When the percent of cells are more than 20% the diagnosis is positive (Dohner et al., 2010). Such a diagnosis of AML can help the above patient to understand the treatment options and appropriate counseling (Betters, 2015).

## 4.5 OVERVIEW OF APPLICATIONS IN A CLINIC

The use of specific antibodies as well as fluorescent tags for viewing specific cells and counting them has made the field of flow cytometry enter studies of immunology as well as hematology apart from preclinical and clinical analysis.

### 4.5.1 Immunology

- An important application of flow cytometry is the analysis of specific subsets of immune cells to look at the probability of the success of a transplant such as bone marrow transplantation. (Jaye et al, 2012)
- The analysis and classification of immune deficiencies can be analysed using the technique. For example: severe combined immune deficiency, autoimmune lymphoproliferative syndrome (Jaye et al, 2012)
- The presence of a population of CD 34 cells can indicate conditions of cardiovascular disease (Subramaniyam et al, 2009).

### 4.5.2 Cancer Diagnosis or Monitoring

The disease has seen much contribution from flow cytometry in prognosis or following up patients with cancer as well as immune studies. The immunophenotype as well as the DNA content can be analysed using flow cytometry which in turn can guide the appropriate therapy. For instance, based on the CD cluster, malignancies that involve CD52 are targeted using alemtuzumab that is anti-CD 52 while B cell cancers with CD20 are treated with anti-CD20 such as rituximab.

Another application is the differentiation between an aggressive mantle cell lymphoma that is also CD11c as compared to B cell cancers that are CD20 to facilitate the appropriate treatment. Thus, a subtle difference between cells with different profiles of antigens can be studied (Kraus et al, 2010).

Another approach is minimal residual disease monitoring for the analysis of malignant cells that remain after treatment (Jaye et al, 2012)

### 4.5.3 Cell Therapy

CD 34 was discovered as an indicator of human hematopoietic stem cells; hence its use was applied in flow cytometry to isolate such cells for transplantation. Such an approach was applied for cancer patients who have been subjected to chemotherapy and radiation as these stem cells can replenish depleted cells. Such approaches have seen positive results as shown by the following studies:

*0.5 X $10^6$ CD34 and CD90low positive were injected per kg into 26 non-Hodgkin's lymphoma reported a reduction of tumor cells by 3.3–5.7 log along with a long term 70% survival over 4 years (Vose et al, 2001)
*The transplantation of 1.1 X $10^6$ CD34 positive and CD90low positive cells per kg into 22 breast cancer patients reported a long-term outcome of 23% patients alive after a decade (Muller et al, 2012)

### 4.5.4 Transplantation

The field of transplantations or transfusions can benefit from the studies of HLA molecule antibodies. Initial studies showed that only flow cytometry could detect such HLA antibodies that are associated with rejection of grafts. The use of microparticles that had been coated with HLA antigens (Class I or II) were found to give results with more sensitivity as compared to whole cells (Jaye et al, 2012).

Further advancements have been realized with the use of cytometry-based multiplexing platform: such as that reported by Taylor et al, 2001 that use fluorescent microparticles (up to 100) to be displayed as an array. Each bead is identified by a unique fluorescence pattern where based on the binding of a fluorescent probe to the bead the signal intensity is reflective of amount of binding. Such developments have allowed the way organs are provided to such patients in countries such as the United States. This application can be extended for genotyping HLA. As it is sensitive to report differences in an amino acid between HLA molecules such genotyping can be accomplished by flow cytometers (Jaye et al, 2012).

## 4.6 FLOW CYTOMETRY AND HEMATOLOGY

There are several flow cytometers of today that offer advantages of cost and an increase in user friendliness. Several specimens such as blood, urine, cerebrospinal fluid, fluids of bone marrow and even tissues can be analyzed. With the blood system having a heterogeneous and distributive nature, flow cytometry of this system has been explored. Several types of cells and their surface markers have been explored such as those of platelets, WBCs and RBCs. A few applications are enumerated here.

### 4.6.1 Analysis of platelets

When antibodies bind to platelets, flow cytometry can be applied for detection of platelet-bound antibodies as well as free antibodies. A fluorescent dye thiazole orange can bind RNA and can thus bind reticulated or immature platelets whose numbers as enumerated flow cytometry can indicate thrombopoiesis. Such a count of platelets is important as several issues with platelets or their glycoproteins.

### 4.6.2 Analysis of RBCs

Paroxysmal nocturnal hemoglobinuria (PNH) can be traced back to a deficient linker on the surface of RBC that can manifest as hemolysis within vessels. It is diagnosed by sugar water test and the Ham's acid hemolysis test that require stringent conditions for testing. Here flow cytometry comes to the rescue with the use of antibodies to CD55 and CD59.

In the event of fetomaternal hemorrhage that arises due to incompatibility of Rh factor of RBC between a fetus and mother, mostly the test called Kleihauer-Betke acid-elution is applied that faces challenges of involving time and labor not to mention variability between observers and labs. More use of flow cytometry is reported where fluorescently labeled antibodies against D antigen (rhesus) are employed that reports of higher sensitivity and reproducibility. The following figure is a scheme of flow cytometry in such a scenario: The peak on the left are RBCs in adults that lack hemoglobin F. F cells are the adult cells bearing hemoglobin F while fetal cells have larger amounts of hemoglobin F that can be detected by antibodies bound to fluorescent tags. Thus, adult and fetal cells can be differentiated.

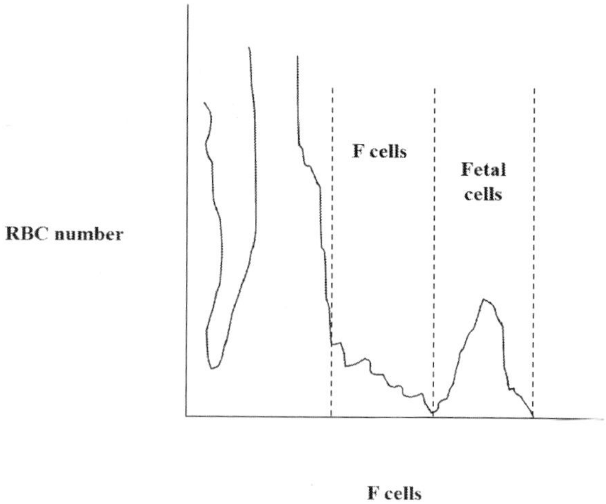


**Figure 4.2:** Flow cytometry for hemoglobin (a schematic reconstruction).

Blood smears are used to evaluate reticulocytes that require staining and manual counting. An automated option is the use of flow cytometry to count the numbers of cells using dyes that bind RNA present in immature cells. This aids in distinguishing between immature and mature cells. Such counts are used in assessment of erythropoiesis for anemia studies and engraftment of bone marrows.

### 4.6.3 Analysis of WBCs

Any deficiencies in functions of leukocytes are facilitated by flow cytometry. Here several aspects such as structure, phagocytosis, and oxidative burst are allowed. Deficiency of leukocyte b2 integrin receptor complex (CD11/CD18) leads to of leukocyte adhesion deficiency syndrome type I seen as impaired migration of neutrophils and monocytes. The use of flow cytometry can permit identification of such complexes and hence in diagnosis.

With HIV targeting the CD4 antigen of T lymphocytes, as the disease advances there is a decrease in CD4-positive T lymphocytes whose numbers are indicative of the stage and monitoring of the ailment. Here again flow cytometry enumerates the CD4 cells that can cast light on diagnosis and therapy of patients.

The diagnosis of B-cell lymphoproliferative disorders can be facilitated by the use of antibodies as these diseases have patterns of antigen expression. For example: Hairy cell leukemia shows presence of CD20, CD11c, CD103 while Follicle center lymphoma shows CD10, CD19, CD20 and CD11c.

Several characters can be analyzed together for the purpose of immunophenotyping of malignancies of the blood system such as leukemias and lymphomas. As different of these diseases have different antigen profiles, the use of newer antibodies and gating can facilitate the analysis by flow cytometry. For instance, the diagnosis of acute myeloid leukemias is challenging using phenotype while the use of CD45 with side scatter can aid in identification of phenotype. A few such applications in hematology are summarized in table 4.3.

**Table 4.3:** few measured analytes by flow cytometry

| Measured antigen/ analyte | Application |
|---|---|
| IgG, IgM | Anti-platelet antibodies<br>Histocompatibility cross-matching |
| CD4, CD8 | Study of immunodeficiency |

| | |
|---|---|
| CD34 | Counting of Hematopoietic progenitor cells |
| Rhesus D | Analyze Feto-maternal hemorrhage |
| Leukocyte surface antigens | Phenotyping of malignancies |
| CD3 | Rejection of transplants |

(Reviewed by Brown and Wittwer, 2000)

## 4.7 DNA CONTENT ANALYSIS BY FLOW CYTOMETRY

Flow cytometry was initially applied in the clinic to look at the content of DNA in cells. For instance an early study in 1983 revealed that malignant cells had an increased amount of DNA up to 67% as compared to normal cells. Thus, a connection was made between abnormalities in DNA and prognosis (Barlogie et al., 1983). Malignant cells have abnormal chromosomes or altered numbers of chromosomes that can be quantifies using flow cytometry as compared to normal cells that are diploid (Betters, 2015).

For this purpose, fluorescent dyes that can intercalate into DNA can be used for example propidium iodide. Based on the quantity of DNA in the cell, the amount of fluorescence and hence the signal and content of DNA can be assessed. The aneuploidy or abnormal contents such as gain or loss can be assessed that again can be traced back to malignancy though careful diagnosis is important (Hedley et al, 1993) as such abnormalities can be seen in benign conditions such as adenomas of thyroid gland (Joensuu et al, 1986).

The main edge of using flow cytometry is the quick analysis of a larger number of cells that can then be analysed. The analysis of this aneuploidy can be linked to better survival in the case of multiple myeloma, neuroblastoma or childhood acute lymphoblastic leukemia. While hypodiploid cells in ALL forecast poor prognosis the presence of hyperdiploid cells in such samples offer more hope.

## 4.8 FLUOROCHROMES USED FOR DIAGNOSIS

An advantage of flow cytometry is the ability to scatter light along with fluorescence that has been attached to monoclonal antibodies targeted to cell populations. The simultaneous development of monoclonal antibodies and flow cytometry together have allowed for such changes in the clinical scenario. Such antibodies are directed against a cluster of differentiation' (CD) antigen. Certain antigens are named differently such as, FMC7, a CD20 epitope on B cells that can be used for lymphoproliferative disease. Several antibodies have been developed that are specific for antigens present in the cytoplasm or surface of cells that have been linked to fluorochromes for the purpose of diagnosis.

For example, propidium iodide or 7-- amino-actinomycin D are dyes that bind DNA that aid in the study of viability of cells or apoptosis as these dyes are taken in by dead cells . Such viability studies are essential in the event of outstation samples as well as before engraftment of cells such as CD 34 stem cells. Few dyes used in clinical analysis are summarized in the Table 4.4.

**Table 4.4:** Dyes in flow cytometry for diagnosis

| Dye | Excitation/ Emission wavelength (nm) | Color |
|---|---|---|
| PE, phycoerythrin | 488/ 578 | Yellow |
| FTIC, fluorescein isothiocyanate | 488/ 519 | Green |
| APC, allophycocyanin | 630/660 | Red |
| PerCP, peridinin-chlorophyll-protein | 488 /678 | Red |

The above wavelengths are seen when they are excited by a laser to emit at a higher wavelength. For the fluorochromes PE, FITC and PerCP 488nm is the excitation wavelength hence one laser can be used but to yield different emission (Virgo and Gibbs, 2012)

## 4.9 QUALITY CONTROL FOR CLINICAL DIAGNOSIS OR STUDIES

As we are speaking of flow cytometry in a clinical scenario, it is vital for clinical diagnosis. Depending on the assay, the use of clinical flow cytometry requires the application of internal quality control (IQC) as well as external quality assessment (EQA). Few pointers include:

- The instrument calibration should be ensured in terms of fluidics, optics and the electronics. For such purposes beads are available for example, Calibrite™ beads (BD Biosciences, San Jose, CA, USA).

- For testing the monoclonal antibodies several antigen negative and antigen positive samples are assayed. There are ready to use preparations available for this purpose.
- An excess of antibody can be applied: this makes sure that a lack of antibody doesn't make the result of weak binding or improper results.
- The sample dilution should be adjusted properly using correct cell counts.
- Compensation is accounted for overlap of spectra and antibody titre for new protocols and assays or for any modifications introduced by the manufacturers of the antibody and instrument.
- The maintenance of the instrument is preferred on service that makes regular servicing and maintenance.

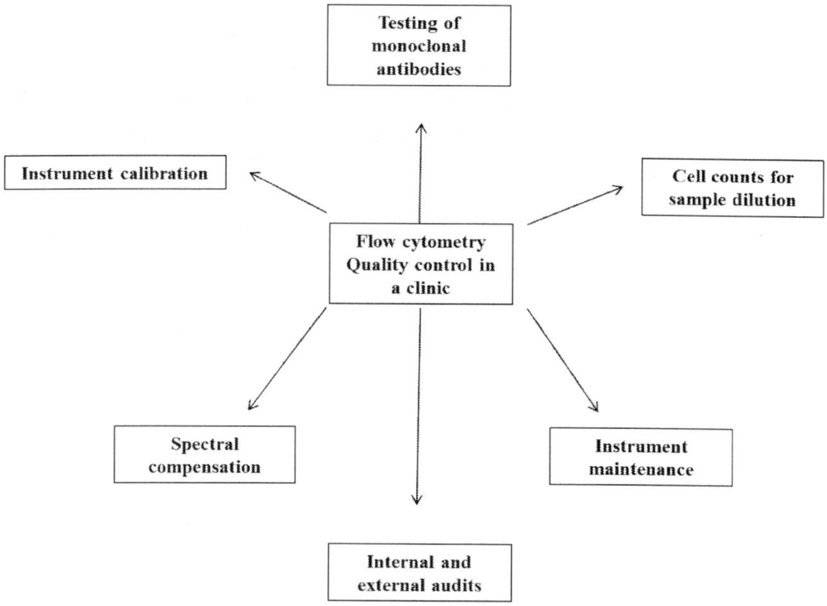


**Figure 4.3:** Overview of quality control.

Such IQC are required to be maintained as records with regular updates of staff training. Documents are required for standard operating procedures. External quality assessment (EQA) in the UK requires Clinical Pathology Accreditation (UK) Ltd (CPA). There are performances reports issued that offer updates and support in the event of any discrepancies or difficulties (Virgo and Gibbs, 2012)

# 4.10 SUMMARY

The conjunction of other techniques such as examination of morphology with flow cytometry can aid pathologists or technicians to achieve a more clear analysis. While morphology can be seen by hematology the use of flow cytometry as discussed in the chapter can aid in differentiation between antigens expressed between cell types. For example, the recurrence of diseases like cancer can be detected before changes in morphology. Challenges to the use of flow cytometry in the lab for diagnosis are the requirement of suspensions for analysis, requirement for viable cells for testing and skilled technicians for operation (Betters, 2015).

This chapter discussed several aspects of application of flow cytometry in a clinical scenario with positive results. The diagnosis of diseases has definitely seen a thrust with the use of flow cytometry-based technologies as well as monitoring of patients with immunological diseases or cancers or even transplants. One of the best examples here is the sorting of CD34 cells for administration of these stem cells in patients with malignancies! The development of complex data with automation along with multidimensional data analysis and high through-put all definitely place flow cytometry on the map for clinical analysis (Jaye et al, 2012).

## 4.11 REFERENCES


1. Alberto Alvarez-barrientos, Javier Arroyo, Rafael Canto´ N, Ce´sar Nombela, and Miguel Sa´nchez-pe´rez (2000). Applications of Flow Cytometry to Clinical Microbiology. Clinical Microbiology reviews. Vol. 13, No. 2. 167–195.
2. Barlogie, B., Raber, M. N., Schumann, J., Johnson, T. S., Drewinko, B., Swartzendruber, D. E, Freireich, E. J. (1983). Flow cytometry in clinical cancer research. *Cancer Research, 43*, 3982–3997.
3. Betters DM (2015) Use of Flow Cytometry in Clinical Practice. J Adv Pract Oncol 2015;6:435–440.
4. David L. Jaye, Robert A. Bray, Howard M. Gebel, Wayne A. C. Harris, Edmund K. Waller. Translational Applications of Flow Cytometry in Clinical Practice The Journal of Immunology May 15, 2012, 188 (10) 4715-4719.
5. Döhner, H., Estey, E. H., Amadori, S., et al (2010). Diagnosis and management of acute myeloid leukemia in adults: Recommendations from an international expert panel, on behalf of the European LeukemiaNet. *Blood, 115*, 453– 474.
6. Hedley DW, Shankey TV, Wheeless LL. DNA cytometry consensus conference. Cytometry 1993;14:471
7. Joensuu H, Klemi P, Eerola E. DNA aneuploidy in follicular adenomas of the thyroid gland. Am J Pathol 1986;124:373–6.
8. Kraus, T. S., C.N. Sillings, D. F. Saxe, S. Y. Li, andD. L. Jaye. 2010. The role of CD11c expression in the diagnosis of mantle cell lymphoma. Am. J. Clin. Pathol. 134: 271–277.
9. Michael Brown and Carl Wittwer. Flow Cytometry: Principles and Clinical Applications in Hematology. *Clinical Chemistry* 46:8(B) 1221–1229 (2000).
10. Mu¨ller, A. M., H. E. Kohrt, S. Cha, G. Laport, J. Klein, A. E. Guardino, L. J. Johnston, K. E. Stockerl-Goldstein, E. Hanania, C. Juttner, et al. 2012. Long term outcome of patients with metastatic breast cancer treated with high-dose chemotherapy and transplantation of purified autologous hematopoietic stem cells. Biol. Blood Marrow Transplant. 18: 125–133.
11. Paul F Virgo and Graham J Gibbs Flow cytometry in clinical pathology. Ann Clin Biochem 2012; 49: 17–28.

12. Subramaniyam, V., E. K. Waller, J. R. Murrow, A. Manatunga, S. Lonial, K. Kasirajan, D. Sutcliffe, W. Harris, W. R. Taylor, R. W. Alexander, and A. A. Quyyumi. 2009. Bone marrow mobilization with granulocyte macrophage colony-stimulating factor improves endothelial dysfunction and exercise capacity in patients with peripheral arterial disease. Am. Heart J. 158: 53–60.

13. Taylor, J. D., D. Briley,Q.Nguyen, K. Long,M.A. Iannone,M.S. Li, F. Ye, A.Afshari, E. Lai, M. Wagner, et al. 2001. Flow cytometric platform for high-throughput single nucleotide polymorphism analysis. Biotechniques 30: 661–666, 668–669.

14. Vose, J. M., P. J. Bierman, J. C. Lynch, K. Atkinson, C. Juttner, C. E. Hanania, G. Bociek, and J. O. Armitage. 2001. Transplantation of highly purified CD341Thy-11 hematopoietic stem cells in patients with recurrent indolent non- Hodgkin's lymphoma. Biol. Blood Marrow Transplant. 7: 680–687.

# CHAPTER 5

# FUNCTIONAL VIRTUAL FLOW CYTOMETRY: A VISUAL ANALYTIC APPROACH FOR CHARACTERIZING SINGLE-CELL GENE EXPRESSION PATTERNS

**Zhi Han[1,2] , Travis Johnson[2] , Jie Zhang[2,3] , Xuan Zhang[1] , and Kun Huang[2]**

[1]College of Software, Nankai University, Tianjin, China

[2]Department of Biomedical Informatics, The Ohio State University, Columbus, OH, USA

[3]The CCC Biomedical Informatics Shared Resource, The Ohio State University, Columbus, OH, USA

**Citation:** Zhi Han, Travis Johnson, Jie Zhang, Xuan Zhang, and Kun Huang, "Functional Virtual Flow Cytometry: A Visual Analytic Approach for Characterizing Single-Cell Gene Expression Patterns," BioMed Research International, vol. 2017, Article ID 3035481, 9 pages, 2017. https://doi.org/10.1155/2017/3035481.

## ABSTRACT

We presented a novel workflow for detecting distribution patterns in cell populations based on single-cell transcriptome study. With the fast adoption of single-cell analysis, a challenge to researchers is how to effectively extract gene features to meaningfully separate the cell population. Considering that coexpressed genes are often functionally or structurally related and the number of coexpressed modules is much smaller than the number of genes, our workflow uses gene coexpression modules as features instead of individual genes. Thus, when the coexpressed modules are summarized into eigengenes, not only can we interactively explore the distribution of cells but also we can promptly interpret the gene features. The interactive visualization is aided by a novel application of spatial statistical analysis to the scatter plots using a clustering index parameter. This parameter helps to highlight interesting 2D patterns in the scatter plot matrix (SPLOM). We demonstrated the effectiveness of the workflow using two large single-cell studies. In the Allen Brain scRNA-seq dataset, the visual analytics suggested a new hypothesis such as the involvement of glutamate metabolism in the separation of the brain cells. In a large glioblastoma study, a sample with a unique cell migration related signature was identified.

## BACKGROUND

Single-cell RNA sequencing (scRNA-seq) is becoming a powerful tool for studying heterogeneity and subtypes in cell populations. Many bioinformatics and computational tools have been developed to visualize, cluster, and categorize the cells based on their expression profiles [1, 2]. Different algorithmic approaches such as principal component analysis (PCA) or multidimensional scaling (MDS) [3], nonnegative matrix factorization [4], minimum spanning tree (MST) [5, 6], latent variable modeling [7], diffusion map [8, 9], and spline models [10] have all been applied and implemented for such purposes. Moreover, it has been shown that often the cells in a population do not always form "clusters." Instead, the cells form a continuous distribution over the space of featured genes and gene signatures [1]. Therefore, it is often of great interest to identify the interesting distribution patterns (e.g., wishbone pattern and bifurcation) which often imply important biological processes such as stem cell differentiation as well as the gene signatures that can be used to reveal such patterns.

However, this effort often leads to a "chicken-and-egg" situation. Since the patterns may not always be readily perceivable from whole genome data, methods such as PCA and MDS may not always be effective. Therefore, it often ends up in an iterative process and a subjective selection of genes of interests. Another commonly adopted workflow is to first cluster the cells based on their expression profiles and identify "gene signatures" that differentiate the clusters followed by enrichment analysis on these signature genes for potential biological functions or processes involved in the separation of the cells. Since there could be many genes involved in differential analysis, the functional enrichment signals can be diluted.

In this paper, we propose a visual analytic workflow called functional virtual flow cytometry (FVFC) for identifying functional gene groups that can effectively separate the cells using scRNA-seq data. We specifically take advantage of gene coexpression network analysis (GCNA). GCNA aims to identify modules of genes with similar expression profiles. It has been well known that the coexpressed genes often are functionally or structurally related [11–16]. Therefore, instead of surveying all the genes, by focusing on the coexpressed gene clusters, we can directly study the cells based on functional gene groups with increased statistical power [17].

Our method is innovative in the following ways. First, it focuses on the gene modules with clear functional relationships (coexpression) and thus greatly enhances the statistical power. Secondly, only the gene modules that are "informative" among the single cells are used. Specifically we focus on the modules that show bimodal or multimodal distributions among the cells to ensure separation power of the genes on the cell population. Thirdly, we apply spatial statistical methods to detect combinations of gene modules that lead to interesting spatial patterns or separation of the cells and thus identify the gene signatures associated with the underlying biological processes. Last but not least, instead of developing this workflow as an "algorithm," we implement it as a visual analytic workflow, allowing the researchers to interactively select gene modules and cell distribution patterns of interest for further investigation. To this end, we take advantage of the SPLOM combined with various visual cues derived from spatial statistical calculation. We demonstrate our workflow using two large single-cell studies on brain and cancer, respectively.

# METHODS

## Workflow

Figure 1 outlines the workflow of our approach that contains three stages. Given a set of processed scRNA-seq data, the first stage carries out the coexpression network analysis and summarization of each network module into a single "eigengene" as well as enrichment analysis to determine the function or structural relationships for each module. The second stage analyzes each eigengene to select the ones with more information content, in particular, the bimodal ones. Then scatterplots are generated for every pair of informative eigengenes. The scatterplots are further analyzed using spatial statistical parameters to determine if they form interesting patterns, specifically if there is clustering or clumping in the scatterplot, implying potential relationships between the two gene modules associated with the two eigengenes. In the final stage, the scatterplots are colored based on the spatial statistical parameters and interesting patterns are further examined with their functional relevance. Overall, this workflow provides an intuitive visual analytic approach for researchers to quickly explore the relationships among functional gene groups in single-cell populations. The details of the steps in the workflow are discussed in the following sections.

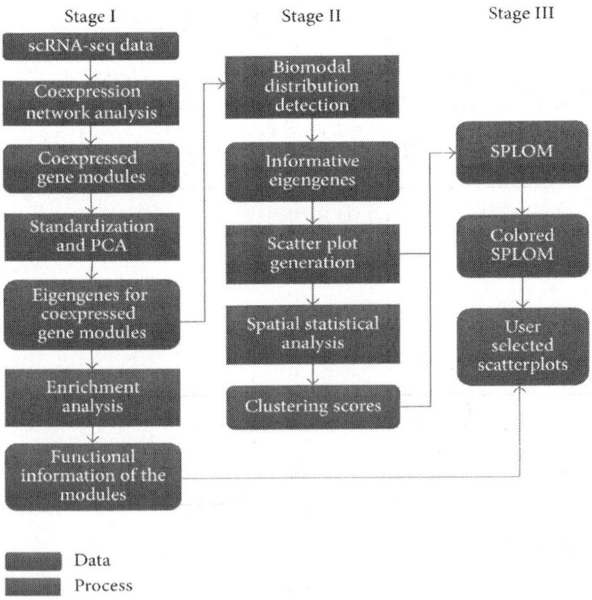


**Figure 1:** The workflow of the functional virtual flow cytometry system.

## Weighted Gene Coexpression Network Analysis

The first stage in Figure 1 is to carry out gene coexpression network analysis. The detailed workflow for this stage is illustrated in Figure 2. Given a set of M genes and their expression levels over N cells, the gene expression profile can be expressed with a matrix

$$\mathbf{G} = \begin{bmatrix} g_{11} & \cdots & g_{1N} \\ \vdots & \ddots & \vdots \\ g_{M1} & \cdots & g_{MN} \end{bmatrix} = \begin{bmatrix} \mathbf{g}_1 \\ \vdots \\ \mathbf{g}_M \end{bmatrix}, \quad (1)$$

where the $N$-dimensional row vector $gi = [g_{i1} \cdots g_{iN}]$ is the expression profile for the $i$th gene across the samples ($i = 1, 2, \ldots, N$). Then the pairwise correlation matrix $C$ can be represented by

$$C = \begin{bmatrix} c_{11} & \cdots & c_{1M} \\ \vdots & \ddots & \vdots \\ c_{M1} & \cdots & c_{MM} \end{bmatrix}, \quad (2)$$

where $c_{ij}$ is the correlation coefficient between $i$th gene vector $g_i$ and $j$th gene vector $g_j$. In our experiment, we use Spearman rank correlation coefficients in the pairwise correlation matrix since Gaussian distribution cannot be assumed for RNA-seq data as required by Pearson correlation.

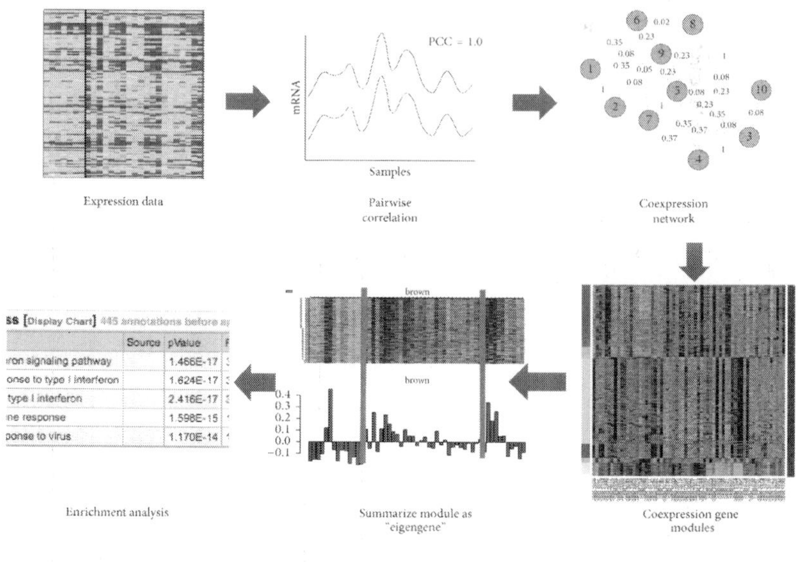


**Figure 2:** Workflow for weighted GCNA and eigengene calculation.

After the correlation matrix was computed, we apply a recently developed algorithm called Normalized lmQCM [15]. Compared to widely adopted gene coexpression network analysis software package WGCNA [18], this algorithm takes a network mining approach allowing overlaps between modules and also is guaranteed to have a lower bound on the density of the detected modules. The output of algorithm lmQCM is a set of gene modules $M_1, M_2,..., M_L$, where each module $M_k = \{i_1, i_2,...,i_{Nk}\}$ is composed of a group of $N_k$ coexpressed genes. The number of modules $L$ and the sizes of the modules are determined by the four parameters of the lmQCM algorithm. While detailed choice of parameters was discussed in [15], the most important parameter is $\gamma$, which is the threshold for the weight of the first edge of any module and thus controls the number of modules. Usually we choose $\gamma$ to ensure that the maximum size of a module is not too large (i.e., less than 500 genes). In addition, we focus on gene modules with at least 10 genes so that meaningful functional enrichment analysis can be applied.

For each gene module detected by lmQCM, $M_k$ can be represented by a gene expression matrix. If we want to compare one gene module against another, it is advantageous to take only a representative of that module rather than taking all the genes. We use PCA to reduce the gene module data meaningfully and take the first principal component as a summary of that module. This first principal component is called "eigengene" in this context. Computationally, we take the submatrix of G for $M_k$ as

$$\mathbf{G_k} = \begin{bmatrix} \mathbf{g_{i_1}} \\ \vdots \\ \mathbf{g_{i_{N_k}}} \end{bmatrix} \in \Re^{N_k \times N}. \quad (3)$$

$G_k$ is centralized and standardized as $G'_k$ such that for each row the mean is zero and the norm is one. Let $G'_k = USV^T$ be the singular value decomposition of $G'_k$. Then the first column of V (denoted as $v_1$) is the "eigengene" for $M_k$ up to a sign since V is an orthonormal matrix whose determinant is 1 or −1. Since the eigengene $w_k$ should reflect the directions of the majority of genes in $G_k$, its projection on the majority of the genes should be positive. Thus, if $\sum \text{sgn}(G_k v_1) < 0$, then $w_k = -v_1$; otherwise, $w_k = v_1$. So each gene module detected by lmQCM corresponds to one "eigengene."

For the reported modules, enrichment analyses are carried out using NIH DAVID (https://david.ncifcrf.gov/) [19] and TOPPGene (https://toppgene.cchmc.org/enrichment.jsp) [20].

## Identify Eigengenes with Bimodal or Long Tail Distribution

Before exploring pairwise relationships between gene modules with eigengenes, we identify and keep eigengenes which are "informative," that is, eigengenes whose distribution follows a bimodal or long tail distribution. Therefore, eigengenes with unimodal distribution, especially the ones with narrow sharp peak-shaped distribution, will be filtered out. To differentiate unimodal distribution with bimodal or long tail distributions, metrics such as Kurtosis, second central differences, and likelihood ratio are adopted [21–23]. Specifically, Kurtosis is a measure of the "tailedness" of the probability distribution of a real-valued random variable [24]. Here we use Kurtosis as a measure to filter whether the histogram of a given eigengene has a very narrow sharp peak distribution. For each eigengene vector $w_k$, first the histogram of the vector is computed and then Kurtosis of the histogram distribution is computed as

$$\mathrm{Kurt}\left(\mathbf{w}_{\mathrm{k}}\right) = \frac{E\left[\left(\mathbf{w}_{\mathrm{k}} - \mu\right)^{4}\right]}{\left(E\left[\left(\mathbf{w}_{\mathrm{k}} - \mu\right)^{2}\right]\right)^{2}}, \tag{4}$$

where $\mu$ is mean of $w_k$. In [24], the Kurtosis value between 3 and 9 show peakness of the distribution while higher values imply sharper peak-shaped distribution. In this paper, we set the threshold for Kurtosis as user defined parameter. If Kurtosis value of histogram for a given eigengene is smaller than a given threshold, then eigengene will be kept.

## Spatial Statistical Analysis of the 2D Scatterplot Using the Nearest Neighbor Distribution

In order to find the relationship between two coexpressed gene modules, we generate pairwise scatter plots for all pairs of eigengene vectors in a 2D space. For two given eigengene vectors $e_i = [e_{i1}, e_{i2},...,]$ and $e_j = [e_{j1}, e_{j2},...,e_{jN}]$, scatter plot is the points with coordinates $(e_{i1}, e_{j1}), (e_{i2}, e_{j2}), \ldots, (e_{iN}, e_{jN})$ in the 2D space. Then we use the nearest neighbor distance (NND) to analyze the pattern. NND for a data point is the distance to its closest neighbor. It is a spatial statistical parameter effectively used for detecting cell patterns in the space [25, 26]. Define $d0$ as the mean NND for all the points. Then we make 100 random simulations, each time the same number of points is created in the same region covering $(e_{i1}, e_{i1})$, $(e_{i2}, e_{j2})$, . . . , $(e_{iN}, e_{jN})$, and the mean NND is calculated. Assuming that $\overline{d_E}$ is the mean of 100 randomly simulated mean NND and $\overline{\sigma}$ is the standard variation, the $z$-score is calculated as

$$z = \frac{\overline{d_0} - \overline{d_E}}{\overline{\sigma}}. \quad (5)$$

We call the z-score as the clustering index for a scatter plot.

## Layout for Visualization

Once the eigengenes with long tail or bimodal distributions are detected, SPLOM is generated. Each scatterplot is then colored using the color scale based on the clustering index. User can then select plots with interesting patters for further visualization and analysis.

# RESULTS

## Datasets and Preprocessing

We applied the above analysis to two large gene expression single-cell datasets. One dataset is RNA sequencing data of single cells isolated from mouse dorsal lateral geniculate nucleus (dLGN) of the thalamus, which is downloaded from Allen Brian Atlas (ABA) website. This data set includes 1,772 single cells collected from dLGN in adult mouse and transcriptionally profiled with RNA sequencing. The dataset contains transcription readings for 45,772 genes and transcripts. However, since many of the genes have zero readings in most cells, these genes were filtered out; specifically we removed genes with zeros in more than half of the cells. In addition, genes whose mean values are among the lowest 20% and variances are among the lowest 50% were removed. This way, 20,000 genes were retained for further analysis.

Another dataset is from a single-cell study on human glioblastoma. The dataset was downloaded from NCBI Gene Expression Omnibus (GEO) with accession number GSE57872. It contains transcriptomes from 430 single glioblastoma cells isolated from 5 individual tumors and 102 single cells from gliomasphere cells lines generated using SMART-seq [27]. Using the same preprocessing procedure, 5,948 genes were kept for further analysis in this dataset.

## Analysis of the ABA Mouse Brain scRNA-Seq Data

Using the lmQCM algorithm (with $\gamma = 0.75$), 60 coexpressed gene modules with at least five genes are identified. Using a threshold 20 for the Kurtosis metric, seven eigengenes are selected. Table 1 summarizes the information for the seven modules. Figure 3 shows the colored SPLOM for the seven eigengenes.

**Table 1:** The seven gene modules whose eigengenes show long tail distributions

| Eigengene # | Index | Size | Kurtosis | Enrichment/notes |
|---|---|---|---|---|
| 1 | 3 | 38 | 10.7844 | 32 predicted genes: three genes are immunoglobulins and two are T cell receptors, acute lymphocytic leukemia ($p = 3.157e - 7$) |
| 2 | 6 | 35 | 5.0379 | Ion transport ($p = 3.341e - 7$), synapse ($p = 2.590e - 7$) |
| 3 | 12 | 18 | 8.5550 | Glutamate decarboxylation to succinate ($p = 7.715e - 7$), inhibitory synapse ($p = 7.843e - 7$) |
| 4 | 13 | 17 | 19.9492 | Development of lower uro neuro e15.5 BladdPelvicGanglion Sox10 top-relative-expression-ranked 1000 ($1.227e - 7$), six genes on chromosome X |
| 5 | 28 | 11 | 4.9068 | Hydrogen ion transmembrane transport ($p = 4.859e - 20$), mitochondrial inner membrane ($p = 1.533e - 16$) |
| 6 | 48 | 6 | 3.8686 | NADH metabolic process ($p = 2.960e - 13$), myelin sheath ($p = 1.643e - 3$), gluconeogenesis ($p = 5.401e - 14$), genes upregulated in hippocampus at late postnatal stages ($p = 9.341e - 10$) |
| 7 | 60 | 5 | 12.5680 | Mostly predicted genes |

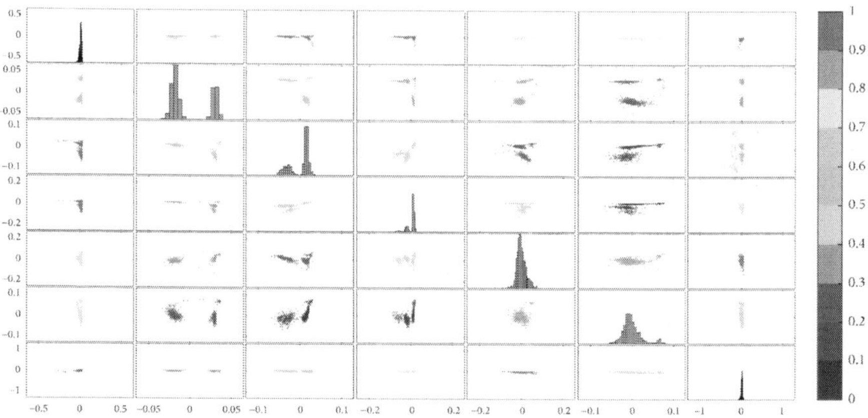

**Figure 3:** Colored SPLOM for the seven long tail eigengenes from the Allen Brain scRNA-seq data. The subplot in the ith row, jth column of the matrix is a scatter plot of the ith eigengene against the jth eigengene. Along the diagonal are histogram plots of each eigengene.

The color scheme in Figure 3 allows us to further inspect scatterplots with interesting patterns. In order to determine if these patterns are associated with specific annotations, for selected scatter plots, we further overlay the annotation information using different colors. Figure 4 shows examples when the broad subtype information about the neurons is overlaid on the scatter plots as points with different colors. It is apparent that none of the gene modules can thoroughly separate the cells based on the subtypes. Instead, some of them can separate specific subtypes. For instance, as in Figure 4(a), the cells are separated into two major clusters based on the "clustering index" as defined in the previous section, which does not fully reflect the subtypes as the blue and yellow points are not separated. Instead, the blue and yellow points are segmented in Figure 4(b) and even further away in Figure 4(c).

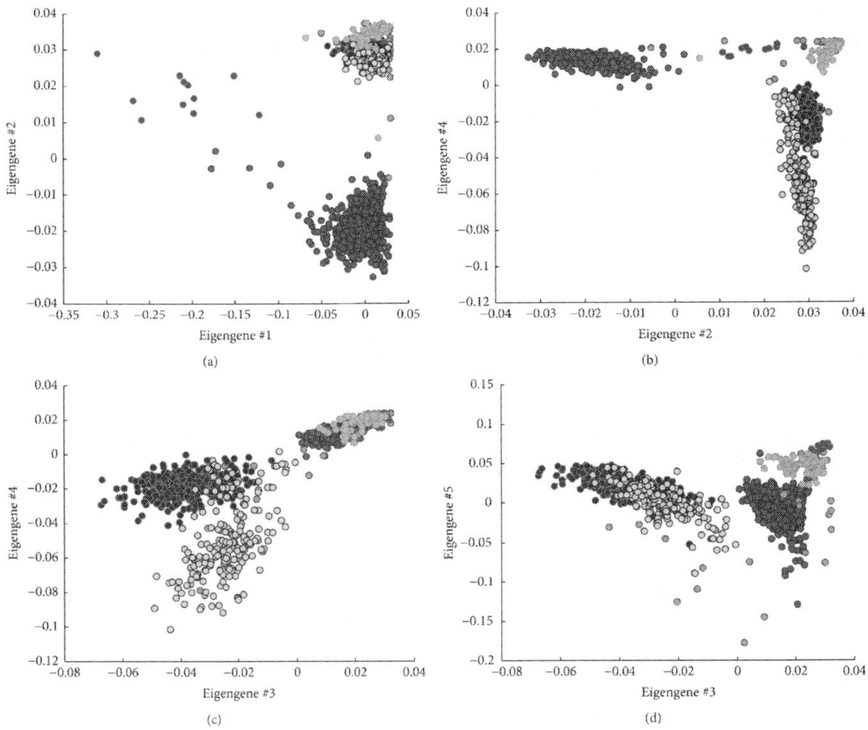


**Figure 4:** Four example scatterplots with broad classes annotated in different colors.

As in Figure 4(b), it is clear that the groups of yellow cells and cyan cells are separated from the rest groups based on eigengene #4 that is enriched with genes that are important to bladder/pelvic ganglion development and may be involved in gender development too. At the same time, it can be noted that the red group is different from the blue, cyan, and yellow groups based on eigengene #2 that is closely associated with synapse formation. In addition, according to Figure 4(c), the blue and yellow groups are separated when both eigengenes #3 and #4 are involved and eigengene #3 is closely connected with the glutamate metabolism and inhibitory synapse development. These neural functions are critical for the interpretation of the cell population clustering.

It is important to notice that the visual outcome is very different from traditional PCA based visualization. As shown in Figure 5(a), if all the genes are used for visualization of the cells using traditional PCA, there is not a clear separation of the cells except for a small group. If we limit the gene features for PCA to the ones involved only in the gene modules listed in Table 1, we can clearly see three major groups. As a control, we marked the three groups of cells in Figure 5(a) with three different colors, and we can see that there is no clear separation of the cells in Figure 5(a). However, without explicit functional grouping, it is difficult to determine which biological processes and functions are involved in such separation.

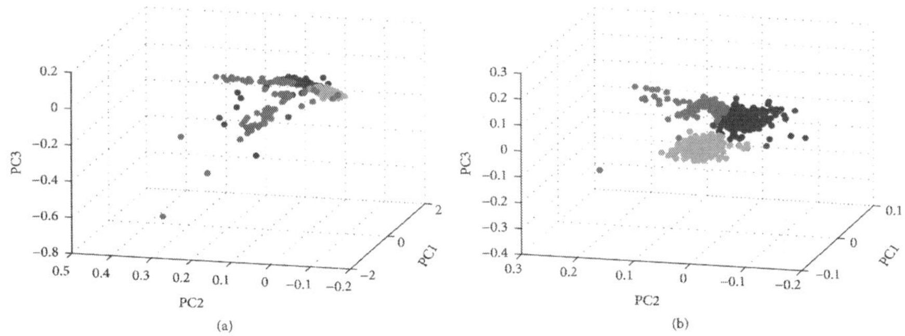


Figure 5: (a) The 3D plot for the first three principal components using all genes for the cells. (b) The 3D plot for the first three principal components using genes in the gene modules in Table 1 for the cells.

## Analysis of the Human Glioblastoma Patients' Brain scRNA-Seq Data

Using the lmQCM algorithm (with $\gamma = 0.2$), 18 coexpressed gene modules with at least five genes are identified. Using a threshold of 5 for the Kurtosis metric, 16 eigengenes are selected.

Figure 6 is the SPLOM for the long tail eigengenes from the brain tumor study.

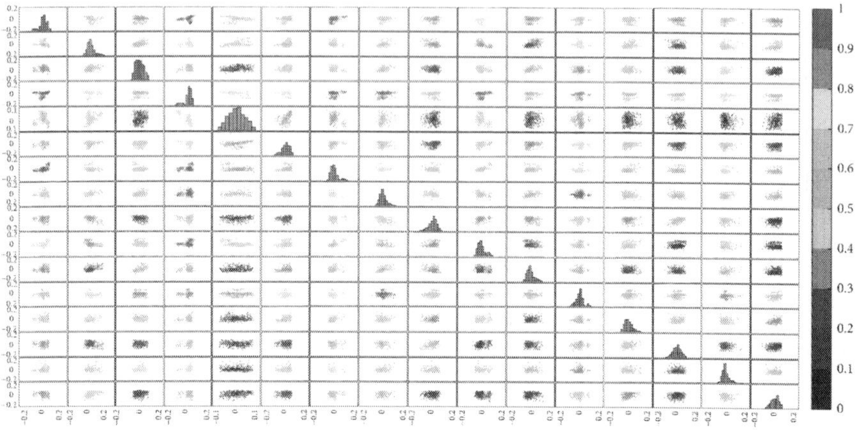

**Figure 6:** Colored SPLOM for the long tail eigengenes from the brain tumor study. The subplot in the ith row, jth column of the matrix is a scatter plot of the ith eigengene against the jth eigengene. Along the diagonal are histogram plots of each eigengene.

From the SPLOM, it is notable that the fourth gene module not only has an eigengene with bimodal distribution but also is involved in effective separation of the cells. While the cells are labeled by the patient and sample IDs, it is clear that some of the separation cases are closely related to the differences between different tumor samples as shown in Figure 7. In particular, eigengene #4 is key in separating the cells in the green group from the rest while other eigengenes can separate other groups (e.g., eigengene #6 separates the yellow cell group from the rest while eigengene #11 separates the red cell group). Interestingly enrichment analysis shows that this gene module for eigengene #4 is highly enriched with extracellular matrix genes (14 genes out of 36, $p = 6.304e - 8$) and the cell migration process (10 genes, $p = 9.145e - 5$), suggesting a particular property of the cells in the green group and it is important as the extracellular matrix and cell migration process is considered critical to the invasion of glioblastoma [28, 29].

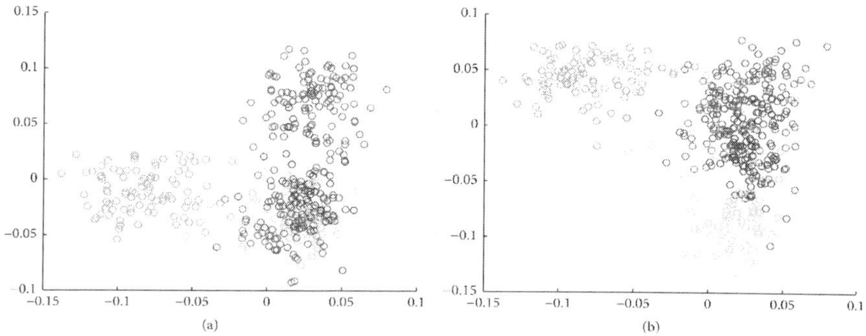


**Figure 7:** (a) The scatter plot between eigengene #4 (x-axis) and eigengene #11 (y-axis). (b) The scatter plot between eigengenes #4 (x-axis) and #6 (y-axis).

# DISCUSSION AND CONCLUSION

In this work, we presented a workflow for detecting distribution patterns in cell populations based on single-cell transcriptome study. With the fast adoption of single-cell analysis, a challenge to researchers is how to effectively extract gene features to meaningfully separate the cell population. However, this often ends up in a chicken-and-egg situation as the separation of the cells often depends on the choice of gene features, yet without a clear pattern it is difficult to determine which gene features are effective. Our workflow uses the well-developed gene coexpression network analysis to take advantage of the fact that coexpressed genes are often functionally or structurally related and the number of coexpressed modules is much smaller than the number of genes. Thus, when the coexpressed modules are summarized into eigengenes, not only can we quickly explore the distribution of cells interactively but also we can promptly interpret the gene features and generate new hypothesis.

Since the cells are separated based on different choices of the gene features, we dub the workflow as “functional virtual flow cytometry,” which achieves separation of the cells based on salient gene features. The separation of cells leads to new hypothesis such as the involvement of glutamate metabolism in the separation of the brain cells in the Allen Brain scRNA-seq data and the specific glioblastoma sample with unique cell migration related signature. While for the latter it is unclear if this observation is indeed biological or due to batch effect, our workflow quickly pointed out the pattern for researchers in deeper examination.

With the interactive visualization, additional advanced analysis can be carried out. For instance, in both Figures 4(b) and 7(b), an interesting observation is that the x- and y-axes cannot both have low values, suggesting interesting Boolean relationships between the gene groups [30]. Therefore, as our ongoing work, these analytic tools along with the workflow are being implemented in an online single-cell analytics portal.

## ACKNOWLEDGMENTS

This work is partially supported by Human Frontier Science Program (to Kun Huang), the NCI ITCR U01CA188547 (to Kun Huang), and the National Natural Science Foundation of China (61572265 to Zhi Han). The Ohio Supercomputer Center provided computing support.

## REFERENCES


1. M. Setty, M. D. Tadmor, S. Reich-Zeliger et al., "Wishbone identifies bifurcating developmental trajectories from single-cell data," Nature Biotechnology, vol. 34, no. 6, pp. 637–645, 2016.
2. S. Krishnaswamy, M. H. Spitzer, M. Mingueneau et al., "Conditional density-based analysis of T cell signaling in single-cell data," Science, vol. 346, no. 6213, Article ID 1250689, 2014.
3. A. Scialdone, K. N. Natarajan, L. R. Saraiva et al., "Computational assignment of cell-cycle stage from single-cell transcriptome data," Methods, vol. 85, pp. 54–61, 2015.
4. C. Shao and T. Höfer, "Robust classification of single-cell transcriptome data by nonnegative matrix factorization," Bioinformatics, vol. 33, no. 2, Article ID btw607, pp. 235–242, 2017.
5. B. Anchang, T. D. P. Hart, S. C. Bendall et al., "Visualization and cellular hierarchy inference of single-cell data using SPADE," Nature Protocols, vol. 11, no. 7, pp. 1264–1279, 2016.
6. P. Qiu, E. F. Simonds, S. C. Bendall et al., "Extracting a cellular hierarchy from high-dimensional cytometry data with SPADE," Nature Biotechnology, vol. 29, no. 10, pp. 886–893, 2011.
7. F. Buettner, K. N. Natarajan, F. P. Casale et al., "Computational analysis of cell-to-cell heterogeneity in single-cell RNA-sequencing data reveals hidden subpopulations of cells," Nature Biotechnology, vol. 33, no. 2, pp. 155–160, 2015.
8. L. Haghverdi, F. Buettner, and F. J. Theis, "Diffusion maps for high-dimensional single-cell analysis of differentiation data," Bioinformatics, vol. 31, no. 18, pp. 2989–2998, 2015.
9. P. Angerer, L. Haghverdi, M. Büttner, F. J. Theis, C. Marr, and F. Buettner, "Destiny: diffusion maps for large-scale single-cell data in R," Bioinformatics, vol. 32, no. 8, pp. 1241–1243, 2016.
10. J. A. DiGiuseppe, M. D. Tadmor, and D. Pe'Er, "Detection of minimal residual disease in B lymphoblastic leukemia using viSNE," Cytometry Part B - Clinical Cytometry, vol. 88, no. 5, pp. 294–304, 2015.
11. B. Zhang and S. Horvath, "A general framework for weighted gene co-expression network analysis," Statistical Applications in Genetics and Molecular Biology, vol. 4, article 17, 2005.

12. A. M. Yip and S. Horvath, "Gene network interconnectedness and the generalized topological overlap measure," BMC Bioinformatics, vol. 8, article 22, 2007.
13. J. Zhang, K. Lu, Y. Xiang et al., "Weighted frequent gene co-expression network mining to identify genes involved in genome stability," PLoS Computational Biology, vol. 8, no. 8, Article ID e1002656, 2012.
14. Z. Han, J. Zhang, G. Sun, G. Liu, and K. Huang, "A matrix rank based concordance index for evaluating and detecting conditional specific co-expressed gene modules," BMC Genomics, vol. 17, article no. 519, 2016.
15. J. Zhang and K. Huang, "Normalized lmQCM: an algorithm for detecting weak quasi-cliques in weighted graph with applications in gene co-expression module discovery in cancers," Cancer Informatics, vol. 1, article 1, p. 137.
16. Y. Xiang, J. Zhang, and K. Huang, "Mining the tissue-tissue gene co-expression network for tumor microenvironment study and biomarker prediction," BMC genomics, vol. 14, supplement 5, article s4, 2013.
17. P. Langfelder and S. Horvath, "Eigengene networks for studying the relationships between co-expression modules," BMC systems biology, vol. 1, article 54, 2007.
18. P. Langfelder and S. Horvath, "WGCNA: an R package for weighted correlation network analysis," BMC Bioinformatics, vol. 9, article 559, 2008.
19. D. W. Huang, B. T. Sherman, and R. A. Lempicki, "Systematic and integrative analysis of large gene lists using DAVID bioinformatics resources," Nature Protocols, vol. 4, no. 1, pp. 44–57, 2009.
20. J. Chen, E. E. Bardes, B. J. Aronow, and A. G. Jegga, "ToppGene Suite for gene list enrichment analysis and candidate gene prioritization," Nucleic Acids Research, vol. 37, no. 2, pp. W305–W311, 2009.
21. A. L. Muratov and O. Y. Gnedin, "Modeling the metallicity distribution of globular clusters," Astrophysical Journal, vol. 718, no. 2, pp. 1266–1288, 2010.
22. J. B. HALDANE, "Simple tests for bimodality and bitangentiality," Annals of Eugenics, vol. 16, no. 1, pp. 359–364, 1951.
23. H. Holzmann and S. Vollmer, "A likelihood ratio test for bimodality in two-component mixtures with application to regional income

distribution in the EU," AStA. Advances in Statistical Analysis, vol. 92, no. 1, pp. 57–69, 2008.

24. L. T. DeCarlo, "On the meaning and use of kurtosis," Psychological Methods, vol. 2, no. 3, pp. 292–307, 1997.
25. K. N. Brown, S. Chen, Z. Han et al., "Clonal production and organization of inhibitory interneurons in the neocortex," Science, vol. 334, no. 6055, pp. 480–486, 2011.
26. H.-T. Xu, Z. Han, P. Gao et al., "Distinct lineage-dependent structural and functional organization of the hippocampus," Cell, vol. 157, no. 7, pp. 1552–1564, 2014.
27. A. P. Patel, I. Tirosh, J. J. Trombetta et al., "Single-cell RNA-seq highlights intratumoral heterogeneity in primary glioblastoma," Science, vol. 344, no. 6190, pp. 1396–1401, 2014.
28. E. T. Sayegh, G. Kaur, O. Bloch, and A. T. Parsa, "Systematic review of protein biomarkers of invasive behavior in glioblastoma," Molecular Neurobiology, vol. 49, no. 3, pp. 1212–1244, 2014.
29. S. M. Turaga and J. D. Lathia, "Adhering towards tumorigenicity: altered adhesion mechanisms in glioblastoma cancer stem cells," CNS Oncology, vol. 5, no. 4, pp. 251–259, 2016.
30. D. Sahoo, D. L. Dill, A. J. Gentles, R. Tibshirani, and S. K. Plevritis, "Boolean implication networks derived from large scale, whole genome microarray datasets," Genome Biology, vol. 9, no. 10, article no. R157, 2008.

# CHAPTER 6

# DETECTION OF LOW ABUNDANCE RNA MOLECULES IN INDIVIDUAL CELLS BY FLOW CYTOMETRY

**Mary Beth Hanley, Woodrow Lomas, Dev Mittar, Vernon Maino, Emily Park**

Research and Development, BD Biosciences, San Jose, California, United States of America

## ABSTRACT

A variety of RNA analysis technologies are available for the detection of RNA transcripts from bulk cell populations. However, the techniques for RNA detection from individual cells have been limited. Here we adapt a novel *in situ* signal amplification method (the RNAScope® detection platform) for the analysis of intracellular RNAs in individual cells by flow cytometry. Using novel target-specific probes that were designed to suppress background signals, we demonstrate the specific detection of HIV gag RNAs in HIV-infected cellular samples, in addition to bcr and abl mRNAs

**Citation:** Hanley MB, Lomas W, Mittar D, Maino V, Park E (2013) Detection of Low Abundance RNA Molecules in Individual Cells by Flow Cytometry. PLoS ONE 8(2): e57002. https://doi.org/10.1371/journal.pone.0057002

in the K562 cell line. This method was capable of distinguishing cells expressing low abundance RNA transcripts and correlated well with quantitative imaging analysis. Furthermore, multiple distinct RNA targets were simultaneously detected with a high specificity without interference. Overall, the sensitivity and specificity of this method will be useful for the analysis of functionally important RNA species from individual cells, even at very low copy numbers.

## INTRODUCTION

Microarrays and quantitative PCR are powerful tools for gene expression analysis that have facilitated our understanding of the intricate biology of normal and disease-state cells and tissues [1]–[4]. Moreover, with the recent advances in high-throughput sequencing technologies, transcriptome profiling by RNA-seq delivers comprehensive gene expression analysis with a large dynamic range [5], [6]. The NanoString® Technologies' nCounter gene expression system reports to have similar sensitivity and accuracy as real-time PCR and includes multiplexing capabilities [7]. These technologies provide the ability to understand the function of genes of interest and also to identify gene expression signatures that distinguish altered biological events from normal events. However, most gene expression studies have used bulk measurements from heterogeneous cells and tissues, in which information from rare or specific cell types can be obscured. By analyzing gene expression in individual cells, a more complete picture of the gene expression dynamics within heterogeneous samples can be captured [8]–[11].

Many single cell analysis tools have been developed and are increasingly applied to address these complex questions [12]–[17], each with its own limitations. Recently RNA-Seq and Fluidigm technologies introduced methods utilizing next generation sequencing or a PCR-based approach allowing for gene expression analysis in single cells, however, these methods require that single cells be isolated prior to analysis[18]–[20]. Flow cytometry, on the other hand, allows for simultaneous measurements of many biomarkers in individual cells in bulk populations. However, such analysis has been limited primarily to proteins and total DNA or highly abundant DNA sequences [21]. Although fluorescent *in situ* hybridization (FISH) technologies have been attempted for high-throughput intracellular RNA analysis by flow cytometry [22]–[24], only limited applications such as acute viral infection or cellular markers with abundant RNA expression have been demonstrated. Since most gene transcripts are present in low quantities (less than 50 copies per cell) [25], the specificity and sensitivity of these RNA FISH technologies are inadequate for the analysis of a broad range of specific gene expression patterns in individual cells.

Recently, a modified form of branched DNA technology has been developed which allows for the visualization of single RNA molecules in cells by image cytometry [26], [27]. Based on the same probe design approach, Wang et al recently reported a novel *in situ* hybridization technology platform (RNAScope) to analyze individual RNA molecules in formalin-fixed, paraffin-embedded tissues [27]. The unique probe design contains paired target probes that form a z-design platform for sequential hybridization-mediated signal amplifications. This strategy allows the visualization of single RNA molecules in cells by simultaneous signal amplification of multiple RNA targets. In addition, the target probe design includes a unique tail sequence type for each probe in the target probe pair that greatly decrease the likelihood of nonspecific hybridizations to occur, yielding superior background suppression.

This report describes the adaptation of RNAScope technology for cell analysis in suspension to enable RNA flow cytometry for single cell analysis of intracellular RNAs in two model systems. The specificity of RNA flow cytometry was confirmed with HIV gag RNA detection in HIV-infected cells. Sensitivity was also established so that the cells expressing specific mRNAs can be separated from the background even when the mRNAs are expressed at very low copy numbers. We also demonstrated the potential utility of multiplexing various RNA targets; using bcr and abl as cellular targets, by multiparametric RNA flow cytometry. The RNA flow cytometry assay described in this article represents a valuable tool for the specific and sensitive detection of multiple RNA transcripts from single cells in heterogeneous biological specimens.

# METHODS

## Ethics Statement

An Institutional Review Board approved informed consent form was used to obtain written informed consent from each blood donor prior to the initiation of these investigational studies.

## Cell Maintenance

Cell lines H9, H9IIIB, and K562 were acquired from the American Type Culture Collection (ATCC). Cells were maintained in I10 medium containing Iscove's Modified Dulbecco's Medium (Life Technologies) supplemented with 10% fetal bovine serum (Life Technologies), and 1% (v/v) penicillin, streptomycin, and L-glutamine (Life Technologies).

Normal whole blood was collected in EDTA BD Vacutainer® collection tubes (BD) through the BD in-house donor program (BD Biosciences, San Jose, CA). Peripheral blood mononuclear cells (PBMCs) were isolated using Histopaque®-1077 (Sigma). For the cryopreservation of PBMCs, cells were washed and resuspended in FBS supplemented with 10% DMSO (EMD Chemicals), the vials placed in a Cryo 1°C freezing container (Nalgene), and subsequently stored at -80°C for 24 hours prior to transfer to liquid nitrogen. Frozen PBMC samples were quickly thawed at 37°C and transferred to 50-ml polypropylene conical tubes, diluted drop-wise in I10 medium, and then centrifuged at 1,500 *g* for 10 minutes.

## Viral Infection of PBMCs

Prior to HIV-1 infection, PBMCs were activated in I10 medium containing 1 μg/ml of phytohemagglutinin (PHA; Sigma). Cells were cultured in PHA-containing medium at $2\times10^6$cells/ml for 2 to 4 days at 37°C with 5% $CO_2$ to allow blasting. PHA-PBMC blasts ($10\times10^6$ cells/ml) were infected with HIV-1 viral stock of NL43 (Advanced Biotechnologies Inc.) at an MOI of 0.01 in I10 medium. Cells were incubated with the virus for 2 hours at 37°C and 5% $CO_2$, at which point I10 medium containing 50 U/ml of interleukin-2 (Roche) was added to bring the cell concentration to 2 $10^6$ cells/ml. A mock infection, in which no virus was added to the blasts, was also set up as a control. Cells were checked at regular intervals for HIVp24 antigen by flow cytometry (per the method described in the following section) to determine when the infected cells were optimal for use in experiments.

## HIVp24 Flow Cytometry

Cells that were intracellular p24 antigen stained were washed with X-Vivo™ 15 medium (Lonza), resuspended in 1% paraformaldehyde (Electron Microscopy Sciences), and incubated for 10 minutes at room temperature. Cells were then centrifuged and resuspended in a permeabilization buffer containing 0.2% saponin (Sigma) in 1X BD FACS™ lysing solution (BD Biosciences) and incubated for 15 minutes at room temperature. After another centrifugation, the cell pellet was resuspended in X-Vivo 15 medium, an anti-p24 PE mAb (BD Biosciences custom conjugation of 37 g12 (Polymun)) was added, and the cell-antibody mixture was incubated at room temperature in the dark for 45 minutes. After washing in X-Vivo 15 medium, the cells were acquired on the BD FACSCalibur™ flow cytometer (BD Biosciences), and analysis was performed using BD FACSDiva™ software or FlowJo™ (Tree Star) software.

## RNA Probe Design

The probes for HIV gag were designed within 670 bases of the p24 coding region based on the consensus sequence among all HIV subtype B sequences for the corresponding region reported to the Los Alamos National Laboratory HIV database by 2009: cctatagtgcagaacctccaggggcaaatggtacatcaggccatatcacctagaactttaaatgcatgggtaaaagtagtagaagagaaggctttcagcccagaagtaatacccatgttttcagcattatcagaaggagccaccccacaagatttaaacaccatgctaaacacagtggggggacatcaagcagccatgcaaatgttaaaagagaccatcaatgaggaagctgcagaatgggatagattgcatccagtgcatgcagggcctattgcaccaggccagatgagagaaccaaggggaagtgacatagcaggaactactagtacccttcaggaacaaataggatggatgacaataatccacctatcccagtaggagaaatctataaaagatggataatcctgggattaaataaaatagtaaggatgtatagccctaccagcattctggacataaaacaaggaccaaaggaaccctttagagactatgtagaccggttctataaaactctaagagccgagcaagcttcacaggaggtaaaaaattggatgacagaaaccttgttggtccaaaatgcgaacccagattgtaagactattttaaaagcattgggaccagcagctacactagaagaaatgatgacagcatgtcagggagtgggaggacc. The probes for the bcr and abl genes were each designed within 1,172 bases of the exon 1 sequence in the bcr gene (accession number NM_004327.3, probed region: nucleotides 779–1850) and 895 bases of sequences spanning multiple exons in the abl gene (accession number NM_005157.4, probed region: nucleotides 88–982). A schematic diagram of the probe region of each gene is shown. Target probes were custom designed for each target gene by Advanced Cell Diagnostics, Inc. (ACD) based on the algorithm described previously[27]. For fluorescence detection, the label probe for the target genes was conjugated to Alexa Fluor® 546 or 647. 18 s rRNA was selected as an internal RNA staining control, and the label probe was conjugated with FITC. All of the reagents, including the target probes and subsequent amplification probes, were manufactured by ACD as a custom order.

## *In situ* RNA Analysis

For the slide based *in situ* RNA analysis, the previously published method was followed [27]. Briefly, cells were fixed on slides and digested with protease, followed by a series of hybridizations with the target-specific probes, amplifiers, and the fluorophore conjugated label probes. The slides were washed thoroughly with Wash Buffer (ACD, proprietary) after each hybridization step. Cells were then counterstained with DAPI (Life Technologies), treated with ProLong™ Gold anti-fade reagent (Life Technologies), and the images were imaged with a 40X or 60X objective on either the BD Pathway™ 435 bioimaging system (BD Biosciences) or an Olympus IX51 microscope containing filter sets from Semrock and Chroma Technology Corp®. Images were captured with a Hamamatsu digital CCD camera. BD AttoVision™ or Cell Profiler (www.cellprofiler.org) software was used for segmentation and image analysis [28], [29].

When immunophenotyping was combined with RNA detection (Fig. 1C), HIV-infected PBMCs were washed with PBS, and incubated with an anti-CD4 mAb conjugated to Alexa Fluor® 488 (BD Biosciences) in PBS. After a 30-minute incubation at room temperature, the cells were washed with PBS and imaged before proceeding with the *in situ* RNA staining procedure beginning with deposition and fixation on slides. Images were taken again after the RNA detection procedure. The two images were overlaid using the DAPI nuclear staining as a guide.

A

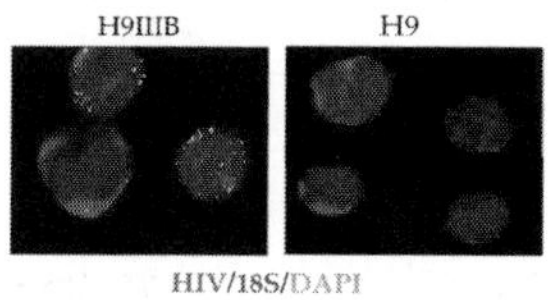


B

| %H9IIIB spiked | 1 | 0.1 | 0.01 | 0 |
|---|---|---|---|---|
| #Total cells | 7,560 | 7,560 | 7,560 | 7,560 |
| #HIV+cells | 114 | 5 | 2 | 0 |
| #HIV gag RNA+ | 1.51 | 0.07 | 0.03 | 0 |

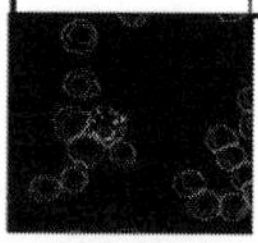

C

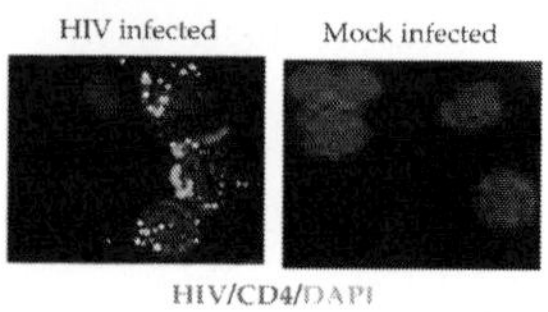


**Figure 1:** Detection of HIV gag RNA *in situ* **in cell lines and PBMCs.**

(a) Representative color merged 60x images of the HIV+ cell line H9IIIB (left) and HIV negative cell line H9 (right); HIV gag RNA-Alexa Fluor® 647 (green), 18 s rRNA-FITC (red), and Nuclei-DAPI (blue). (b) The frequency of HIV gag RNA-positive cells within the 18 s+DAPI+ population was calculated after spiking HIV+ cells into the negative cell population at the stated percentages in the table. The image below the table is representative of data from image segmentation analysis. (c) 40x image of CD4 immunophenotyping overlaid with HIV gag RNA in HIV-infected PBMCs; Anti-CD4 antibody-Alexa Fluor® 488 (red), HIV gag RNA-Alexa Fluor® 546 (green), and DAPI (blue).

https://doi.org/10.1371/journal.pone.0057002.g001

## RNA Flow Cytometry

The same procedure was applied for RNA flow cytometry as was used for the slide based *in situ* RNA detection as described previously, except that the cells were kept in microcentrifuge tubes throughout the staining process in a 100-µl volume for each hybridization step. The cells were centrifuged at 500 *g*, and washes were done in 1 ml of Assay Wash Buffer (ACD, proprietary). Acquisition was completed on either the BD FACSAria™ III (laser 488nm (filter 530/30 with 505 LP filter) for the detection of FITC, laser 561nm (filter 585/15) for the detection of Alexa Fluor®546 and laser 633nm (filter 660/20) for the detection of Alexa Fluor® 647), the BD FACSVerse™ (laser 488nm (filter 527/32 with 507 LP filter) for the detection of FITC and laser 640nm (filter 660/10) for the detection of Alexa Fluor® 647) , the BD FACSCanto II (laser 488nm (filter 530/30 with 502 LP filter) for the detection of FITC and laser 633nm (filter 660/20) for the detection of Alexa Fluor® 647), or the BD LSRFortessa™ (laser 488nm (filter 520/50 with 505 LP filter) for the detection of FITC and Alexa Fluor 488, laser 532nm (filter 582/15) for the detection of Alexa Fluor®546, and laser 640nm (filter 670/30) for the detection of Alexa Fluor® 647) (BD Biosciences). The BD FACSAria III was used for cell sorting. Sorted cells were deposited onto slides for image analysis, as described previously. All of the flow cytometry data analysis was performed using BD FACSDiva (BD Biosciences) or FlowJo (Tree Star) software.

## Statistical Analysis

Graphs and statistical analyses were completed using Microsoft® Excel® software.

# RESULTS

## Specificity of *in situ* RNA Detection

To evaluate the specificity of the RNAScope probes for *in situ* RNA detection, we designed an HIV gag-specific probe. To visualize HIV gag-specific RNA detection, imaging cytometry was assessed. Slides were prepared with the H9 cell line chronically infected with the HIV-IIIB strain (H9IIIB) and with the HIV-uninfected cell line H9. The slides were hybridized with the HIV gag probe, followed by image acquisition and analysis.

HIV gag RNA was detected as distinct fluorescent spots only in the HIV-infected H9IIIB cells, in contrast to the H9 cells, for which no detection of HIV gag RNA was observed (Fig. 1A). Using 18s ribosomal RNA (rRNA) as an internal control (depicted as red in the images in Fig. 1A), we identified and included only live cells and excluded dead cells and debris during image analysis. When the HIV-infected cells were spiked into uninfected cells at various low frequencies (0.01 to 1%) prior to hybridization, a distinct gag RNA staining pattern was visible, as shown in the representative outline image after segmentation (Fig. 1B). A quantification of the HIV RNA+ cells, within the identified 18 s rRNA+DAPI+ cells, yielded the expected frequencies (table in Fig. 1B). HIV RNA signal was not detected in the control slides, where no target probes were used during the staining procedure. This data is representative of six independent experiments.

After demonstrating HIV RNA specificity with precision in cell lines, we next tested primary cells. Peripheral blood mononuclear cells (PBMCs) from normal uninfected individuals were stimulated and expanded for 3 days and then infected with $HIV_{NL43}$ as described in methods. Briefly, PBMC blasts were infected and cultured until p24+ cells were readily detectable (within 3 to 10 days) by intracellular protein flow cytometry for HIVp24. HIV gag RNAs were seen only in the HIV-infected PBMCs and not in the mock-infected PBMCs (Fig. 1C). To confirm the CD4 specificity of HIV infection, cells also were stained with Alexa Fluor® 488 anti-CD4 antibody before the RNA staining procedure. HIV RNA was detected only in CD4-positive subsets of HIV-infected PBMCs, as expected [30], confirming the biological receptor specificity of HIV infection in this T-cell subpopulation (Fig. 1C).

## Comparison of RNA Staining in Cells on Slides versus in Suspension

Once we established the specificity of the RNA staining in cells fixed on slides, we next wanted to determine whether the same RNA staining technology could be applied to cells kept in a suspension. A 1:1 mixture of H9IIIB and H9 cells was prepared and a portion of the cells was deposited and processed on slides. The remainder of the cells was kept in a suspension for the RNA staining procedure. The cells kept in suspension were processed as stated in methods. Briefly, the cells were processed in microcentrifuge tubes for the procedure, and the washes were performed by centrifugation and supernatant aspiration. After the RNA staining procedure, the cells in suspension were deposited onto slides for imaging. Both sets of slides were imaged and analyzed for HIV gag-positive cell frequency and HIV gag signal intensity within nucleated live cells. The punctuate staining pattern in both methods was visually similar, and the percentage of HIV+ cells in both cases was very near the expected 50% (Fig. 2A).

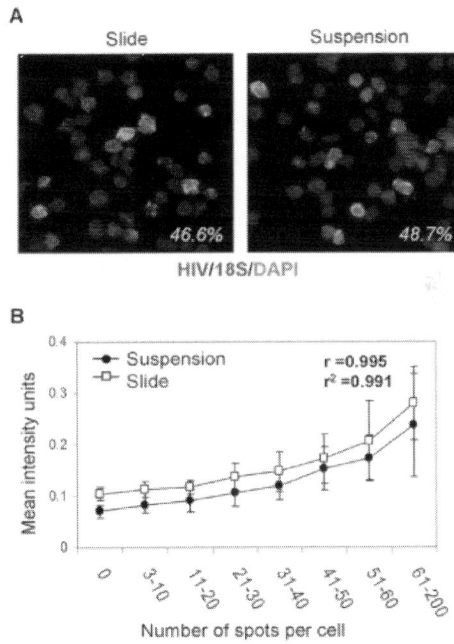


**Figure 2:** Comparison of HIV gag RNA *in situ* **detection by a slide-based method vs a suspension-based method.**

(a) Pseudocolor-merged 40x images of a 1:1 mixture of H9 and H9IIIB cells via the slide-based RNA detection method (left) and the suspension method (right) of HIV gag RNA-Alexa Fluor® 546, 18 s rRNA-FITC, and DAPI. The calculated frequencies for HIV gag+18 s+ RNA for each are depicted in each image. (b) Mean intensity comparison of HIV gag for different HIV gag spot count ranges (bins) with the suspension-based and slide-based methods. Error bars depict the standard deviation within each spot count bin. The correlation coefficient, r, and the coefficient of determination, $r^2$, are shown on the graph.

https://doi.org/10.1371/journal.pone.0057002.g002

The mean fluorescence intensity (MFI) of the HIV RNA signal in the suspension stained cells was slightly lower than that of the cells stained on the slides, though still comparable. Not surprisingly, for both preparation methods, the distribution of the number of spots per cell was highly correlated with the distribution of cellular fluorescence intensity, as evident from the correlation coefficient being close to 1 (Fig. 2B). This data was reproduced and also confirmed using bcr Alexa Fluor® 647 and abl Alexa Fluor® 546 in the K562 cell line.

## Analysis of Intracellular RNAs by Flow Cytometry

Having obtained similar RNA staining results with both the slide and the suspension- based assays using imaging, we next set out to establish whether the stained suspension cells could be acquired on a flow cytometer with a specificity and sensitivity similar to that with imaging. A mixture of H9IIIB and H9 cells was prepared, which contained approximately 20% HIV+ cells, to compare the results from two platforms, imaging and flow cytometry. The mixed cell suspension was stained for HIV gag RNA and 18 s rRNA and then divided into two portions. One portion of the stained cells was deposited onto a slide, and images were acquired. The remaining cells were acquired in suspension on the flow cytometer. Quantitative analysis of the two preparations yielded the anticipated result: 20.7% and 18.9% of the cells were defined as HIV+ by image and flow analysis, respectively (Fig. 3A). A control that omitted target probes during the RNA staining procedure was used to assess nonspecific staining (data not shown) and to set the negative gates for the flow cytometry analysis (Fig. 3A). Similar results were observed when the HIV gag Alexa Fluor® 546 label probe was used in place of Alexa Fluor® 647 in a separate experiment (data not shown).

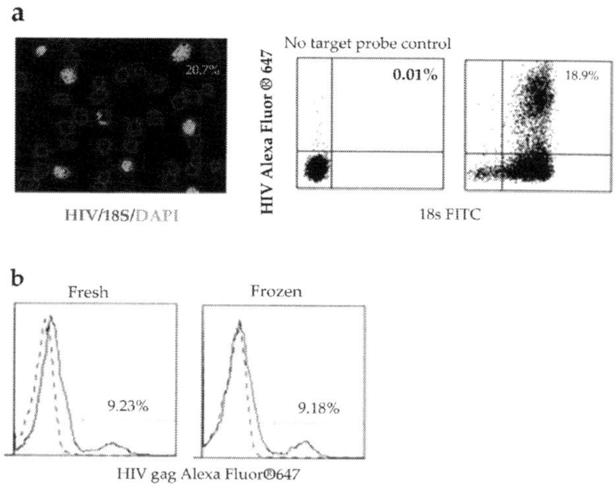


**Figure 3:** Validation of the RNA flow cytometry procedure.

(a) A pseudocolor merged segmentation mask image obtained using Cell Profiler software analysis of a mixture of H9 and H9IIIB cells (left image); HIV gag-Alexa Fluor® 647, 18 s-FITC, and DAPI counterstain. The calculated frequency of HIV gag+ cells (within the 18 s+DAPI+ cells) is shown on the image. The RNA flow cytometry plots of the same mixture of H9 and H9IIIB cells. (b) Flow cytometry overlay histograms of HIV gag RNA in HIV-infected PBMCs (solid line) and mock-infected PBMCs (tinted with dashed line) in freshly acquired PBMCs (left plot) and the same PBMCs after cryopreservation (right plot).

https://doi.org/10.1371/journal.pone.0057002.g003

To further investigate RNA flow cytometry in primary cells, *in vitro* HIV-infected PBMCs were prepared as described in methods. A mock-infected sample was included as a control. Since it is often important to be able to ship primary samples, and to be able to analyze from previously frozen samples for retrospective studies, we also analyzed the effects of cryopreservation on these HIV-infected and mock-infected PBMCs. A significant HIV gag RNA-positive population was seen in the RNA analysis of both fresh and cryopreserved infected samples (solid lines) compared to mock-infected controls (dashed lines) (Fig. 3B, data is representative of eight independent experiments).

## Cellular mRNA Analysis and Multiplex RNA Flow Cytometry

To evaluate the feasibility of RNA flow cytometry beyond the HIV target, we designed RNA target probes for messenger RNAs of the bcr and abl genes. The performance of these target probes was evaluated in the K562 cell line expressing the bcr-abl fusion construct [31]. The bcr and abl targets were tested individually as well as together. Each target was clearly detectable by flow cytometry analysis when individually tested as well as when the target probes were combined.

Utilizing the bcr, abl, and 18 s RNA target probes, we then assessed the multiplexing capability of RNA flow cytometry in the K562 cell line. In the multiplex analysis, a clear population of bcr and abl double-positive cells was identified within the 18 s rRNA+ gated cell population (Fig. 4A). One no-target probe control was used to set the negative gates for all of the targets. When the bcr+abl+18 s+ cells were subsequently sorted and imaged on a slide, closely paired spots likely representing the mRNA of bcr-abl fusion constructs were clearly visible within the cells (yellow in Fig. 4B). The ratio of yellow bcr-abl fusion spots to green abl spots confers with our qRT-PCR data for the ratio of fusion transcripts to abl mRNA quantitation (data not shown). These data suggest that the yellow spots visualized in these slides post sort are indeed bcr-abl fusion RNA, which validates the potential for subcellular analyses of gene expression using the cell sorting function of flow cytometry.

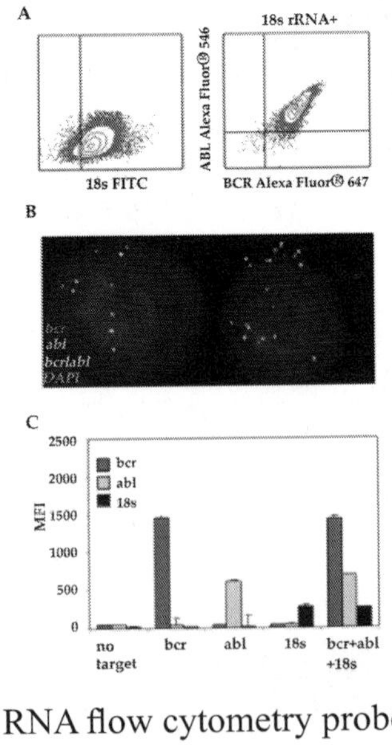


**Figure 4:** Assessment of RNA flow cytometry probe multiplexing.

(a) Flow cytometry plots from three-probe (bcr, abl, 18 s) RNA analysis in the K562 cell line: 18 s rRNA (left) and bcr vs. abl RNA in the 18 s+ events (right). (b) 60x pseudocolored images of sorted bcr+abl+18 s+ cells from K562 cells; Alexa Fluor® 647 labeled bcr (red), Alexa Fluor® 546 labeled abl (green). The bcr/abl fusion transcripts are shown in yellow due to the merging of both the Alexa Fluor® 546 and Alexa Fluor® 647 dyes. Cells were counterstained with DAPI. (c) Graph showing the effects of RNA staining by multiple probes (bcr+abl+18 s) compared to a single target probe on MFIs in RNA flow cytometry. Bars represent the standard deviation for duplicate samples run.

https://doi.org/10.1371/journal.pone.0057002.g004

MFI data generated by flow cytometry analysis of each individual RNA target was compared with a multiplex sample containing all three targets (bcr, abl, and 18 s). We observed no apparent difference in the MFI when a single target probe was added or when all three probes were added (Fig. 4C). Additionally, in the multiplex analysis of bcr, HIV gag, and 18 s RNAs in the HIV-negative K562 cell line, there was no indication of interference when a non-relevant probe (HIV gag) was included with the specific target probes (18 s and bcr) during the hybridization process. Overall, there was no indication of interference among the different probe sets during the signal amplification process in the multiplex RNA flow cytometry analysis.

## Sensitivity of RNA Flow Cytometry

To further evaluate RNA flow cytometry for the analysis of low-abundant intracellular RNA, the detection sensitivity was assessed using bcr as the target RNA in the K562 cell line. Given that RNA staining was comparable between cells on slides and cells in suspension (Fig. 2), we directly compared the MFI data obtained by flow cytometry with the RNA spot counts calculated by image analysis from the same original cell suspension. K562 cells were stained in suspension with target probes for bcr, a portion of the cells was deposited on a slide, and images were acquired. The remaining cell suspension was acquired on the flow cytometer. RNA spots for the bcr target were calculated from image analysis, and spot frequencies were grouped into bins, to segment the data points for analysis (Fig. 5A). The respective percentage of cells within each bin was then applied to the corresponding flow cytometry data. This was done by setting the analysis gates in the flow histogram on the bcr RNA-positive cells (Fig. 5B) such that each gate's frequency was equivalent to the percentage calculated from the image spot count analysis (y-axis values from Fig. 5A). The corresponding MFI for each gate could then be obtained in the flow cytometry analysis (table in Fig. 5B). To assess the detection sensitivity of RNA flow cytometry, the lowest spot count bin (1–5 range) was further split into individual spot counts and the MFIs were determined (inset to the table in Fig. 5B). From this analysis it was determined that the cells containing one or two RNA spots could be distinguished using RNA flow cytometry. Considering that each RNA is detected as an individual spot by the RNAScope slide based assay and that this assay was shown to be a single molecule detection method with a signal detection efficiency of 85% [27], this data indicates that cells containing very few RNA copies are able to be distinguished from the background.

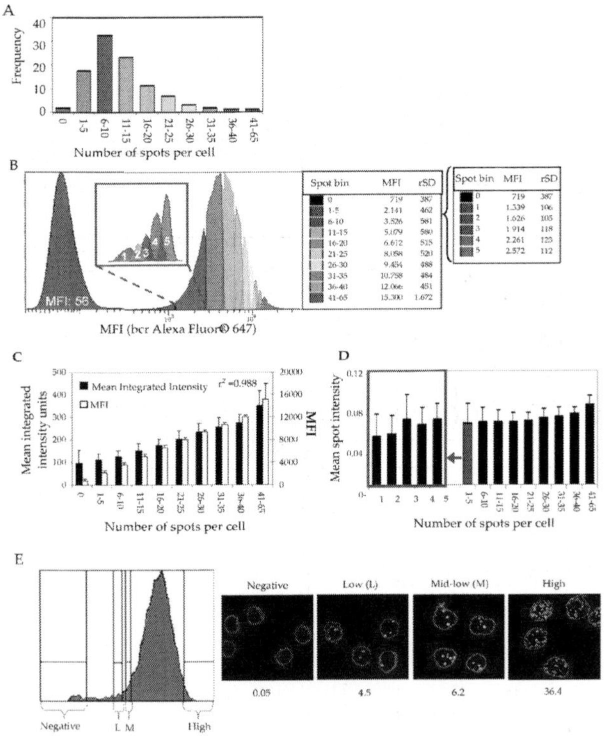


**Figure 5:** Validation of the sensitivity of RNA flow cytometry.

(a) Breakdown of spot numbers from image analysis (bcr+18 s+DAPI+) into bins (x-axis) with corresponding frequencies (y-axis). (b) RNA flow histogram of K562 cells split into groups using the percentages obtained in Figure 5A with the corresponding MFIs (x-axis and table). Colors correspond to the spot bin numbers in the bar chart in 5a. The lowest bin (spot count 1–5) was further subdivided and is shown in the inset on the histogram with the corresponding spot count on each peak. (c) Comparison chart of mean integrated intensities (black bars) obtained from image analysis and MFIs (white bars) obtained from RNA flow cytometry analysis (5b). Standard deviations are shown for each group. The coefficient of determination ($r^2$) is displayed on the graph (0.988). (d) Spot intensities for each spot bin with correlating standard deviations for each group. The lower bin (1–5) was further subdivided and is shown on the left side of the chart. (e) Histogram showing sort gates for the different bcr MFI subsets in the K562 cell line. Outline overlay images shown to the right of the histogram are representative of the sorted cells from each subset sorted showing bcr RNA (green), 18 s rRNA (red), and DAPI counterstaining (blue). The bcr spot count average per cell is listed below its respective image.

https://doi.org/10.1371/journal.pone.0057002.g005

When the MFIs from the flow cytometry analysis (Fig. 5B) were plotted against the mean integrated intensities from the imaging analysis for each bin, we found a very good correlation between the twρ measurements ($r^2$ value of 0.988, Fig. 5C). Additionally, the individual spot intensities were found to be very consistent among the different spot count bins, even at the lower end of the detection limit (Fig. 5D). When we sorted different subsets of bcr RNA-positive cells, based on MFI, and subsequently deposited the sorted cells onto slides for RNA image analysis (Fig. 5E), the spot frequency correlated with the MFI from the flow histogram (Fig. 5E). This provides confirmation that the MFIs obtained in RNA flow cytometry acquisition correspond to the HIV spot numbers and intensities obtained in RNA slide image analysis.

## DISCUSSION

Over the last few decades, flow cytometric detection of intracellular RNAs has been attempted by applying various molecular technologies [15]–[17], [32]. However, the specificity and/or sensitivity of all previous attempts have not been suitable for the wide range of intracellular RNA analyses, particularly in the case of low abundance of target RNA sequences. Here we have described a method for RNA flow cytometry with performance characteristics enabling the simultaneous analysis of multiple RNAs in individual cells at the sensitivity-detection limit for a single RNA molecule. By adapting a novel signal amplification technology [27] that provides superb background suppression with its unique target specific probe design, we could analyze the highly amplified signal of even low frequency specific mRNAs by flow cytometry without increasing the background signal.

We have established the quantitative capabilities of RNA flow cytometry by demonstrating that the signal amplification per each RNA molecule is quantitatively consistent among all cells in a population (Fig. 5D). Furthermore, the imaging cytometry data from RNA amplification on slides correlated with the flow cytometry analysis on suspension cells. The amplified signals were highly specific for each target RNA, as evidenced in the HIV gag model, with little or no signal from the HIV-uninfected cells or cells treated with either no or irrelevant target probes (Fig 3A). Furthermore, we demonstrated the sensitivity of RNA flow cytometry to be applicable for the analysis cells containing very low RNA copies in each cell (Fig. 5B). In this study, up to three targets in multiplex RNA flow cytometry have been validated (Fig. 4A). However, we believe that the level of multiplexing could be further expanded for a higher complexity gene expression analysis, since no sign of inter-target interference during the signal amplification was observed, even when the two target RNAs were in close proximity, as we demonstrated with the bcr-abl fusion transcript (Fig. 4B).

To the best of our knowledge, this is the first demonstration of a sensitive flow cytometric analysis of specific messenger RNA in individual cells. The high sensitivity and specificity of the presented method enable the analysis of RNA molecules present in low copy numbers from individual cells and in minority subpopulations. Additionally, RNA flow cytometry will be extremely valuable in rare cell analyses that involve both dynamic and heterogeneous populations of cells. This single cell RNA analysis method can be applied to various types of cellular samples, including cell lines, primary cells, and even previously cryopreserved cells (Fig. 3B).

We believe that RNA flow cytometry, as demonstrated in this study, represents a novel tool to validate gene expression profiles in individual cells or for a comprehensive evaluation of the expression dynamics of genes from different cell types. It will be particularly valuable for understanding the complex network of pathways in cells among very heterogeneous cell populations and might lead to widespread applications in areas including stem cell biology, oncology, immunology and immune cell-related diseases, and developmental biology. A combined RNA flow cytometry analysis with protein targets, such as the preliminary data shown in Figure 1B, would be a highly desirable future application, in cases when the subsets of cells have already well defined antigen expression.

## ACKNOWLEDGMENTS

We thank Yuling Luo and Nan Su of Advanced Cell Diagnostics, Inc. for technical aid, Robert H. Singer at Albert Einstein College of Medicine for constructive input to the program design, and John Dunne of BD Biosciences for providing helpful comments on the manuscript.

## AUTHOR CONTRIBUTIONS

Conceived and designed the experiments: EP VM. Performed the experiments: MBH WL. Analyzed the data: MBH WL. Contributed reagents/materials/analysis tools: MBH WL. Wrote the paper: MBH EP DM.

# REFERENCES

1. Freeman WM, Robertson DJ, Vrana KE (2000) Fundamentals of DNA hybridization arrays for gene expression analysis. Biotechniques 29: 1042–1046, 1048–1055.
2. Symmans WF, Ayers M, Clark EA, Stec J, Hess KR, et al. (2003) Total RNA yield and microarray gene expression profiles from fine-needle aspiration biopsy and core-needle biopsy samples of breast carcinoma. Cancer 97: 2960–2971.
3. van 't Veer LJ, Dai H, van de Vijver MJ, He YD, Hart AA, et al. (2002) Gene expression profiling predicts clinical outcome of breast cancer. Nature 415: 530–536.
4. VanGuilder HD, Vrana KE, Freeman WM (2008) Twenty-five years of quantitative PCR for gene expression analysis. Biotechniques 44: 619–626.
5. Iacobucci I, Ferrarini A, Sazzini M, Giacomelli E, Lonetti A, et al. (2012) Application of the whole-transcriptome shotgun sequencing approach to the study of Philadelphia-positive acute lymphoblastic leukemia. Blood Cancer J 2: e61.
6. Wang Z, Gerstein M, Snyder M (2009) RNA-Seq: a revolutionary tool for transcriptomics. Nat Rev Genet 10: 57–63.
7. Geiss GK, Bumgarner RE, Birditt B, Dahl T, Dowidar N, et al. (2008) Direct multiplexed measurement of gene expression with color-coded probe pairs. Nat Biotechnol 26: 317–325.
8. Di Carlo D, Tse HT, Gossett DR (2012) Introduction: why analyze single cells? Methods Mol Biol 853: 1–10.
9. Narsinh KH, Sun N, Sanchez-Freire V, Lee AS, Almeida P, et al. (2011) Single cell transcriptional profiling reveals heterogeneity of human induced pluripotent stem cells. J Clin Invest 121: 1217–1221.
10. Raj A, Tyagi S (2010) Detection of individual endogenous RNA transcripts in situ using multiple singly labeled probes. Methods Enzymol 472: 365–386.
11. Raj A, van Oudenaarden A (2008) Nature, nurturc, or chance: stochastic gene expression and its consequences. Cell 135: 216–226.
12. Bengtsson M, Stahlberg A, Rorsman P, Kubista M (2005) Gene expression profiling in single cells from the pancreatic islets of Langerhans reveals lognormal distribution of mRNA levels. Genome

Res 15: 1388–1392.

13. Citri A, Pang ZP, Sudhof TC, Wernig M, Malenka RC (2011) Comprehensive qPCR profiling of gene expression in single neuronal cells. Nat Protoc 7: 118–127.
14. De Spiegelaere W, Filliers M, Van Soom A (2012) Laser capture microdissection for gene expression analysis of specific cell populations in single blastocysts. Methods Mol Biol 853: 29–37.
15. Kalisky T, Quake SR (2011) Single-cell genomics. Nat Methods 8: 311–314.
16. Femino AM, Fay FS, Fogarty K, Singer RH (1998) Visualization of single RNA transcripts in situ. Science 280: 585–590.
17. Levsky JM, Shenoy SM, Pezo RC, Singer RH (2002) Single-cell gene expression profiling. Science 297: 836–840.
18. Hashimshony T, Wagner F, Sher N, Yanai I (2012) CEL-Seq: single-cell RNA-Seq by multiplexed linear amplification. Cell Rep 2: 666–673.
19. Gourronc FA, Robertson M, Herrig AK, Lansdorp PM, Goldman FD, et al. (2010) Proliferative defects in dyskeratosis congenita skin keratinocytes are corrected by expression of the telomerase reverse transcriptase, TERT, or by activation of endogenous telomerase through expression of papillomavirus E6/E7 or the telomerase RNA component, TERC. Exp Dermatol 19: 279–288.
20. Spurgeon SL, Jones RC, Ramakrishnan R (2008) High throughput gene expression measurement with real time PCR in a microfluidic dynamic array. PLoS One 3: e1662.
21. Herzenberg LA, Parks D, Sahaf B, Perez O, Roederer M, et al. (2002) The history and future of the fluorescence activated cell sorter and flow cytometry: a view from Stanford. Clin Chem 48: 1819–1827.
22. Robertson KL, Thach DC (2009) LNA flow-FISH: a flow cytometry-fluorescence in situ hybridization method to detect messenger RNA using locked nucleic acid probes. Anal Biochem 390: 109–114.
23. Robertson KL, Verhoeven AB, Thach DC, Chang EL (2010) Monitoring viral RNA in infected cells with LNA flow-FISH. RNA 16: 1679–1685.
24. Crouch J, Leitenberg D, Smith BR, Howe JG (1997) Epstein-Barr virus suspension cell assay using in situ hybridization and flow cytometry. Cytometry 29: 50–57.

25. Zhang L, Zhou W, Velculescu VE, Kern SE, Hruban RH, et al. (1997) Gene expression profiles in normal and cancer cells. Science 276: 1268–1272.
26. Luo Y, Ma Y, Nguyen C (2010) Multiplex detection of nucleic acids. United States: Advanced Cell Diagnostics, Inc.
27. Wang F, Flanagan J, Su N, Wang LC, Bui S, et al. (2012) RNAscope: a novel in situ RNA analysis platform for formalin-fixed, paraffin-embedded tissues. J Mol Diagn 14: 22–29.
28. Carpenter AE, Jones TR, Lamprecht MR, Clarke C, Kang IH, et al. (2006) CellProfiler: image analysis software for identifying and quantifying cell phenotypes. Genome Biol 7: R100.
29. Jones TR, Kang IH, Wheeler DB, Lindquist RA, Papallo A, et al. (2008) CellProfiler Analyst: data exploration and analysis software for complex image-based screens. BMC Bioinformatics 9: 482.
30. Schnittman SM, Psallidopoulos MC, Lane HC, Thompson L, Baseler M, et al. (1989) The reservoir for HIV-1 in human peripheral blood is a T cell that maintains expression of CD4. Science 245: 305–308.
31. Lozzio CB, Lozzio BB (1975) Human chronic myelogenous leukemia cell-line with positive Philadelphia chromosome. Blood 45: 321–334.
32. Itzkovitz S, van Oudenaarden A (2011) Validating transcripts with probes and imaging technology. Nat Methods 8: S12–19.

# CHAPTER 7

# FLOW CYTOMETRY BASED DETECTION AND ISOLATION OF PLASMODIUM FALCIPARUM LIVER STAGES IN VITRO

**Peter C. Dumoulin[1], Stefanie A. Trop[1], Jinxia Ma[1], Hao Zhang[1], Matthew A. Sherman[2], Jelena Levitskaya[1]**

[1] W. Harry Feinstone Department of Molecular Microbiology and Immunology, Bloomberg School of Public Health, Johns Hopkins University, Baltimore, MD, 21205, United States of America

[2] Triangle Research Labs, 6 Davis Drive, Durham, NC, 27709, United States of America

## ABSTRACT

Malaria, the disease caused by *Plasmodium* parasites, remains a major global health burden. The liver stage of *Plasmodium falciparum* infection is a leading target for immunological and pharmacological interventions. Therefore, novel approaches providing specific detection and isolation of live *P. falciparum* exoerythrocytic forms (EEFs) are warranted. Utilizing

---

**Citation:** Dumoulin PC, Trop SA, Ma J, Zhang H, Sherman MA, Levitskaya J (2015) Flow Cytometry Based Detection and Isolation of *Plasmodium falciparum* Liver Stages *In Vitro*. PLoS ONE 10(6): e0129623. https://doi.org/10.1371/journal.pone.0129623

a recently generated parasite strain expressing green fluorescent protein (GFP) we established a method which, allows for detection and isolation of developing live *P. falciparum* liver stages by flow cytometry. Using this technique we compared the susceptibility of five immortalized human hepatocyte cell lines and primary hepatocyte cultures from three donors to infection by *P. falciparum* sporozoites. Here, we show that EEFs can be detected and isolated from *in vitro*infected cultures of the HC-04 cell line and primary human hepatocytes. We confirmed the presence of developing parasites in sorted live human hepatocytes and characterized their morphology by fluorescence microscopy. Finally, we validated the practical applications of our approach by re-examining the importance of host ligand CD81 for hepatocyte infection by *P. falciparum* sporozoites *in vitro* and assessment of the inhibitory activity of anti-sporozoite antibodies. This methodology provides us with the tools to study both, the basic biology of the *P. falciparum* liver stage and the effects of host-derived factors on the development of *P. falciparum* EEFs.

## INTRODUCTION

Infection with *Plasmodium* parasites, the causative agent of malaria, remains a major public health problem. In 2012 an estimated 207 million new cases of malaria occurred resulting in an estimated 627,000 deaths, primarily in sub-Saharan Africa [1]. Of the five currently known human malaria parasites, *Plasmodium falciparum* causes the highest rates of complications and mortality [1]. The *Plasmodium* life cycle in humans consists of two phases: the clinically silent liver stage, or exoerythrocytic form (EEF), and the erythrocytic stage [2, 3]. The latter is routinely studied both *in vitro* [4] using red blood cell cultures and *ex vivo* using patient-derived infected blood [5–7].

Direct access to *in vivo* infected human hepatocytes is untenable due to ethical and logistic constraints. Consequently, studies of the liver stage of *Plasmodium* infection have relied mainly on the use of rodent parasites both *in vivo* and *in vitro* [8, 9]. The rodent parasites *Plasmodium berghei* and *Plasmodium yoelii* complete full development in the hepatocyte in less than three days after infection and can fully develop in human hepatocellular carcinoma cell lines *in vitro*[10, 11]. However, the human parasite *P. falciparum* requires at least 144 hours for full EEF development in the liver and has a limited ability to infect human hepatocellular carcinoma cell lines [9].

Multiple experimental models utilizing primary human hepatocytes for *P. falciparum* EEF development have been reported. Infection of primary hepatocytes *in vitro* by *P. falciparum* was first described almost thirty years ago [12]. Recent work using micropatterned primary hepatocytes surrounded by stromal cells has allowed for both complete development of *P. falciparum* EEFs and possibly generation of *P. vivax* hypnozoites *in vitro* [13]. The first mouse models relying on the engraftment of human hepatocytes into immune-compromised animals capable of generating mature EEFs were reported more than two decades ago [14] and were further used to obtain isolated infected cells from fixed frozen liver tissues through micro-dissection [15]. Complete development of *P. falciparum* liver stages and liver-to-blood transmission was later demonstrated *in vivo* in immune-compromised and fumarylacetoacetate hydrolase-deficient animals backcrossed with NOD mice [16]. Recently, SCID mice with chimeric human livers were used to show the protective effect of parasite antigen-specific human monoclonal antibodies derived from RTS,S vaccine recipients [17].

The *in vivo* and *in vitro* methods described above demonstrated the generation of *P. falciparum*merozoites capable of infecting red blood cells. However, the technical complexity and high associated costs restrict the widespread use of these methodologies for routine studies on *P. falciparum* liver stages. Additionally, these methods rely on immunofluorescence or quantification of total parasite biomass and are unable to isolate live, individual *P.falciparum*EEFs. Therefore, a technically reproducible and cost-effective experimental system for *in vitro*monitoring and purification of *P. falciparum* EEFs is still needed.

Mouse models of the liver stage of malaria infection suggest a role for both $CD8^+$ T cells and sporozoite antigen-specific antibodies in sterilizing immunity [18]. However, understanding the contributions of humoral and cell mediated immune responses directed against *P. falciparum*EEFs during the natural course of infection [19, 20] or induced upon vaccination [21, 22] requires a robust *in vitro* system.

Two modes of interaction between sporozoites and host hepatocytes are currently described *in vitro* [23, 24] and *in vivo* [25, 26]: (i) breaching of the host cell plasma membrane followed by intracellular movement and subsequent exit, referred to as traversal, and (ii) productive invasion and parasitophorous vacuole formation within hepatocytes. The influence of traversed cells on infection and parasite biology are largely unknown. Thus, an optimal experimental *in vitro* system recapitulating the liver stage of *P.*

*falciparum* should allow for specific identification and isolation of traversed from non-traversed and infected from non-infected cells. In experimental models of *Plasmodium* infection *in vitro* non-traversed and non-infected populations are similarly exposed to a plethora of biological factors from the salivary glands of infected mosquitoes. Thus, these populations of hepatocytes serve as the most accurate control to study the immunology and developmental biology of *P. falciparum* liver stage infection *in vitro*. However, *in vivo*, sporozoites are deposited into the host dermis and the direct impact of salivary gland content on hepatocytes is likely to be insignificant [27].

Here we report a flow cytometry based *in vitro* system to monitor *P. falciparum* liver stages that permits (i) detection and isolation of *P. falciparum* EEFs, (ii) evaluation of host factors on the establishment of an exoerythrocytic infection, and (iii) efficacy assessment of antibodies blocking sporozoite motility.

# MATERIALS AND METHODS

## Human Hepatocyte Culture

HC-04 [28] was obtained from the ATCC (Manassas, VA, USA). HepG2 and HepG2-CD81 [29] were kindly provided by Dr. I. Coppens (Johns Hopkins Bloomberg School of Public Health). These cell lines were maintained in IMDM supplemented with 2.5% FCS, 100 IU/ml penicillin, 100 μg/ml streptomycin and 2 mM L-glutamine. THLE-2 and THLE-3 cell lines [30] were kindly provided by Dr. C. Harris (NCI, Center for Cancer Research) and maintained using BEGM BulletKit medium (Lonza, Walkersville, MD) supplemented with 5 ng/ml EGF, 70 ng/ml phosphoethanolamine and 10% FCS. All cell lines were routinely tested for the presence of mycoplasma [31].

Cryopreserved human primary hepatocytes, hepatocyte thawing, plating and maintenance medium were obtained from Triangle Research Labs (TRL, North Carolina). Hepatocytes were thawed in a 37°C water bath for 2 minutes in thawing medium, spun at 100xg for 8 minutes and plated on collagen-coated wells or coverslips in plating medium. Plating medium was replaced with maintenance medium supplemented with 10% FCS 6 hours after plating and replaced every 24 hours.

## Infection with *P. falciparum* Sporozoites

Prior to addition of sporozoites, HC-04, HepG2 and HepG2-CD81 were seeded at densities of 400,000, 200,000 or 100,000 cells/well in 12, 24 or 48 well plates respectively, to achieve confluence. THLE-2 and THLE-3 were seeded at 200,000, 100,000 or 50,000 cells/well in 12, 24 or 48 well plates to achieve confluence. Primary hepatocytes were seeded at 300,000 hepatocytes per well in a 24 well plate 48 hours prior to addition of sporozoites. Sporozoite to hepatocyte ratios are indicated for each experiment in relevant figure legends. *Plasmodium falciparum* 3D7HT-GFP [32] sporozoites were isolated and purified by density gradient as described previously [33]. Sporozoites were added to hepatocytes in IMDM with 2.5% FCS supplemented with penicillin, streptomycin and L-glutamine, centrifuged at 380x*g* for 5 minutes without break and then incubated at 37°C for 4–6 hours. After incubation, cells were washed with PBS and maintained in complete growth medium supplemented with Fungizone (1.25ug/ml, Gibco, Grand Island, NY). To prevent overgrowth of HC-04, HepG2, and HepG2-CD81 cells, each well was individually processed at 24 hours postinfection (pi), split at a 1:2 ratio and re-seeded separately into a new plate. Medium was changed daily. Fungizone treatment was terminated at 48 hours pi.

## Hepatocyte Traversal

Hepatocyte traversal was performed as previously described [23, 24]. At the time of sporozoite addition or mock infection medium was supplemented with 0.2 mg/ml of 10,000 MW dextran-tetramethylrhodamine (Life Technologies, Grand Island, NY). After incubation at 37°C for 6 hours cells were washed three times in PBS, trypsinized and re-suspended in PBS/0.1% BSA. Dextran retention was detected using FACSCalibur and CellQuest acquisition software (Becton Dickinson, New Jersey, USA). Data analyses were performed using FlowJo software (TreeStar Inc., Ashland Oregon, USA).

## Detection of Infected Hepatocytes by Flow Cytometry

Hepatocytes from individual wells were washed three times in PBS, trypsinized, re-suspended in PBS/0.1% BSA and analyzed using a Becton Dickinson LSR II or FACSCalibur (Becton Dickinson, New Jersey, USA). Dead cells were excluded by propidium iodide (PI) staining. Data analyses were performed using FlowJo software (TreeStar Inc., Ashland Oregon, USA). GFP-positive PI-negative events were identified using a ratio of the

FL1 (530/30) and FL2 (585/42) fluorescence channels. For quantification of infected cell populations, percentages of $GFP^{+}$cells as well as geometric mean fluorescent intensity (geoMFI) of the gated GFP positive viable (PI-negative) population are indicated.

## Quantification of Infection by Real-time PCR

Hepatocytes were lysed using the Power SYBR Green Cells-to-$C_T$ kit (Ambion, Grand Island, NY) at the density of 2000 cells/μl. The lysis reaction was stopped with a proportional volume of 10x Stop Solution. Reverse transcription was performed using 22.5 μl of cell lysate in a total volume of 50 μl, and 4 μl of the resulting cDNA were used as template per PCR reaction. PCR amplification used the following primer pairs: Human GAPDH 5′-gcaaattccatggcaccgt-3′/5′-tcgccccacttgattttgg-3′ *P. falciparum* 18S 5′-tcagataccgtcgtaatctta-3′/5›-aactttctcgcttgcgcgaa-3′ [34]. Each sample was amplified in triplicate using an Applied Biosystems StepOnePlus and SYBR Green for detection. Thermal cycling proceeded for 2 minutes at 50°C, 10 minutes at 95°C followed by 40 cycles of 15 seconds at 95°C and 1 minute at 60°C. *Pf*18S copy numbers were calculated using a standard curve generated from serial dilution of a plasmid template containing the *Pf*18S gene. Samples were normalized to their own GAPDH $C_T$ values.

## Immunofluorescence of GFP-positive HC-04 and Primary Hepatocytes

GFP-positive HC-04 cells were sorted 96 hours pi directly into 4% paraformaldehyde using a Beckman Coulter MoFlo Cell Sorter (Johns Hopkins Bloomberg School of Public Health, Flow Cytometry and Cell Sorting Core Facility). Fixed cells were cytospun at 1500 rpm for 5 minutes onto cytoslides (Thermo Shandon, Astmoor, UK). Specimens were blocked and permeabilized in PBS containing 10% goat serum, 1% BSA, and 0.1% Triton X-100, and then stained with anti-*Pf*Hsp70 (cl. 4C9, F. Zavala, JHU) [35] followed by anti-mouse IgG-Alexa Fluor 594 (Molecular Probes, Grand Island, NY). Slides were mounted with ProLong Gold Antifade containing DAPI (Molecular Probes, Eugene, OR) and viewed on a widefield Nikon 90i microscope. Data were acquired and processed with Volocity software (Perkin Elmer, Waltham, MA).

Primary hepatocytes seeded on collagen-coated glass coverslips were infected, fixed and stained with an anti-*Pf*Hsp70 antibody at 96 hours pi as described above. Quantifications of parasite size and distance from host nucleus were performed in Volocity software, using the *Pf*Hsp70-specific fluorescence signal as margins for surface area and longest axis calculations. Minimum distance was calculated from the edge of *Pf*Hsp70 to the edge of the nearest hepatocyte nucleus.

## CD81 Detection and Neutralization

Single cell suspensions of hepatocytes were incubated for 20 minutes on ice with either anti-CD81 (clone JS-81, BD Pharmingen, San Diego, CA) or IgG1 isotype control antibody, each conjugated to APC. Azide-free anti-CD81 mouse monoclonal antibody (cl.1D6) (Abcam, Cambridge, MA) or a functional grade IgG1 isotype control antibody (eBioscience, San Diego, CA) were used in CD81 blocking experiments at a final concentration of 10 μg/ml. Antibodies were either added 2 hours prior to infection and removed, added concurrently with sporozoites, or added at 6 hours pi.

## Evaluation of Efficacy of Mouse Serum Containing Humanized Anti-CSP mAb 2A10

Mouse serum containing humanized CSP-specific monoclonal antibody 2A10 [36–38] (h2A10) was kindly provided by Dr. Gary Ketner and Dr. Cailin Deal (Johns Hopkins Bloomberg School of Public Health) [39]. Briefly, sera samples were collected from a single mouse prior to and 11 weeks after intra-muscular administration of $1x10^{11}$ genome copies of AAV-expressing humanized 2A10 (anti-CSP). Concentration of human IgG in total immune serum was 1.93 mg/ml. Serum were added at the same time as sporozoites.

## Statistical Analysis

Comparisons of multiple groups were performed with a one way ANOVA. A post-hoc Tukey's test was used for comparisons between individual groups. Adjusted p-values are shown where relevant (*p<0.05, **p<0.01, ***p<0.001). There was considerable variability in infectivity between sporozoite preparations; consequently, relevant comparisons are made only between sporozoites from the same preparations or are normalized where relevant.

# RESULTS

## Immortalized Human Hepatocyte Cell Lines are Permissive for Traversal by *P. falciparum* Sporozoites

Traversal of hepatocytes by *Plasmodium* sporozoites has been previously documented *in vitro*[23, 24] and *in vivo* [25, 26]. The *in vitro* traversed cell population was identified by flow cytometry by its ability to retain fluorescent dextran during migration of sporozoites through the host cell [24]. Gating strategy for flow cytometry-based detection of hepatocytes traversed with *P. falciparum* 3D7HT-GFP sporozoites is shown. In search of cell lines that allow for efficient detection of traversal and invasion of *P. falciparum* sporozoites, we turned to immortalized cell lines derived from human hepatocytes that included THLE-2 and THLE-3 [30], HepG2, HepG2-CD81 [29] and HC-04. HC-04 is the only cell line previously reported to support development of *P. falciparum* liver stages [13, 28].

We found that all five cell lines tested had a detectable population of hepatocytes traversed by *P. falciparum* 3D7HT-GFP sporozoites (Fig 1A). HC-04, HepG2 and HepG2-CD81 had comparable percentages of cells retaining fluorescent dextran, which was about 2–3 fold higher than in cultures of THLE-2 and THLE-3 cell lines (Fig 1B). We concluded that each of the tested cell lines can be utilized to study traversal by *P. falciparum* sporozoites *in vitro*. Traversal was measured at 6 hours after addition of sporozoites. At this time we are unable to detect any GFP-positive hepatocytes by flow cytometry. In agreement with previous studies utilizing rodent parasites [24] we found that fluorescent dextran uptake in human hepatocyte cultures incubated with *P. falciparum* sporozoites depends on hepatocyte-to-sporozoite ratio and could be inhibited by preincubation of sporozoites with cytochalasin D, an inhibitor of sporozoite motility. Of note, the sporozoite-to-hepatocyte ratios used in our study are significantly higher than those reported previously [24] and, therefore, higher percentages of traversed cells are detected in our experiments. Purification of sporozoites by density the gradient centrifugation method used in our study removes the majority of mosquito salivary gland material that generates a cleaner and more concentrated preparation.

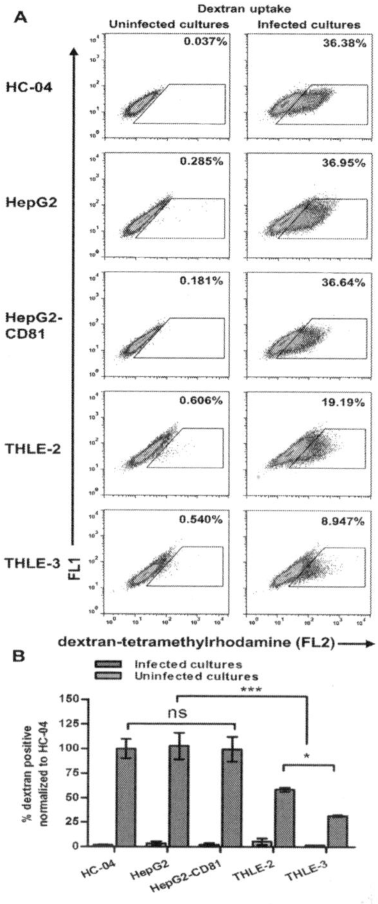


**Figure 1:** Human hepatocyte cell lines can be traversed by *P. falciparum* **sporozoites.**

Traversal was assessed by flow cytometry as an ability of cells to uptake and retain fluorescent high molecular weight dextran during sporozoite co-incubation for 6 hours. (A) Representative plots for each cell line are shown, 1.5:1 sporozoite-to-hepatocyte ratio was used. Numbers indicate percentage of dextran-positive cells. (B) Percentage of traversed cells normalized to HC-04 obtained from three sporozoite preparations, each experiment was conducted in triplicate. Mean +/- SD shown.

https://doi.org/10.1371/journal.pone.0129623.g001

## Human Hepatocyte Cell Lines Exhibit a Differential Ability to Support Development of *P. falciparum* EEFs

Next, to assess the ability of immortalized hepatocyte cell lines to support the development of *P. falciparum* 3D7HT-GFP parasites, we used a novel, quantitative, flow cytometry-based analysis of infected cultures. We utilized relative FL1 (530/30) to FL2 (585/42) fluorescence ratios to detect the weak GFP signal of the parasite in relation to background autofluorescence of the host cell while simultaneously excluding dead cells using propidium iodide. To validate our ability to specifically detect developing parasites we quantified parasite 18S expression using real-time PCR and visualized *P. falciparum* EEFs by immunofluorescence microscopy following flow cytometry-based sorting of infected cells.

Here we show that $GFP^+$ events could be detected not only in HC-04, but also in HepG2 and HepG2-CD81, which had been previously described as non-permissive for the development of *P. falciparum* EEFs (Fig 2A) [9]. However, the relative frequencies (percentages) of $GFP^+$ $PI^-$events were 2.5-fold higher (Fig 2B) and the total number of $GFP^+$ events per well were 8-fold higher (Fig 2C) in HC-04 as compared to HepG2 and HepG2-CD81 cells. Both THLE-2 and THLE-3 had significantly fewer total $GFP^+$ events than the other cell lines tested (Fig 2C). In contrast, the fluorescence intensity of the $GFP^+$ populations, an indication of parasite development, was comparable in all cell lines tested (Fig 2D).

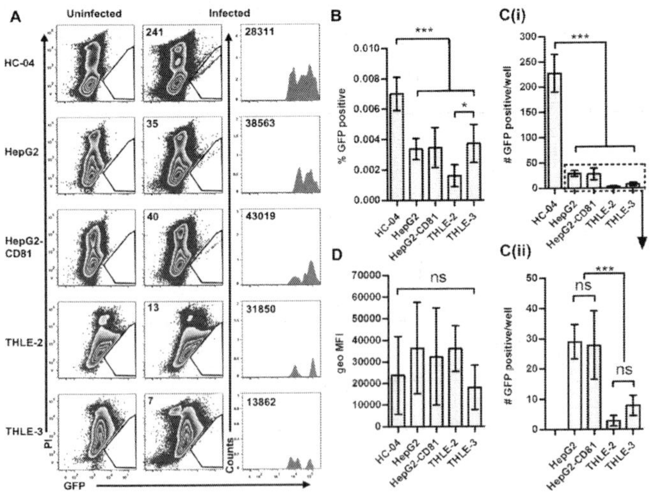


**Figure 2:** Identification of hepatocytes infected with *P. falciparum* **3D7HT-GFP parasites by flow cytometry**.

Sporozoites were added at a 2.3:1 sporozoite-to-hepatocyte ratio. Infected cells were identified by flow cytometry at 96 hours postinfection as GFP-positive events in PI-negative (viable) cell populations. Uninfected cultures propagated in parallel with infected cultures were used to define specificity of GFP-positive events. A similar number of total events were acquired from infected and uninfected cultures of each cell lines. (A) Data from one representative culture is shown for each cell line. Event number corresponding to the GFP-positive gate (middle panels) and the geometric mean of events in the GFP-positive gate is indicated (right panels). (B) Percentage of GFP-positive PI-negative events. (C) Number of GFP events per well (D) Geometric MFI in FL-1 of GFP-positive PI-negative events. Mean +/- SD shown for all bar graphs (n = 6 per cell line).

https://doi.org/10.1371/journal.pone.0129623.g002

Quantification of *P. falciparum* 18S rRNA at two time points postinfection supported our flow cytometry findings (Fig 3). The ratio of parasite 18S copy number relative to a hepatocyte housekeeping gene (GAPDH $C_{T,}$ arbitrary units) was 2-fold higher in HC-04 than in HepG2 and HepG2-CD81, and about 30-fold higher than in THLE-2 and THLE-3 (Fig 3A). In HC-04, a 3-fold increase in the *Pf*18S rRNA copy numbers per well was observed between 48 and 96 hours postinfection (Fig 3B). However, a decrease in the 18S:GAPDH ratio was observed in the same cultures from 48 to 96 hours postinfection (Fig 3A). This can be explained by extensive proliferation of uninfected HC-04 cells (Fig 3C) resulting in the relative drop in percentage of infected hepatocytes.

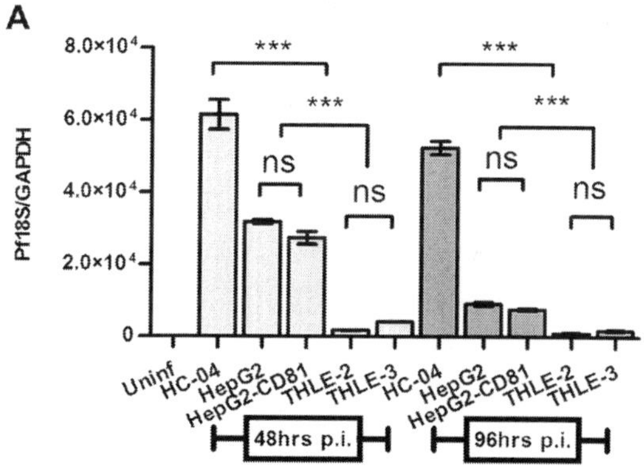

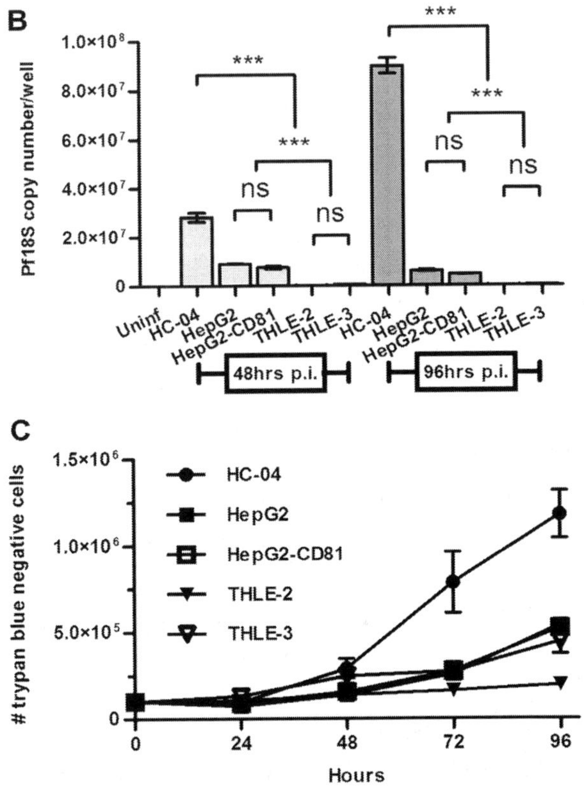


**Figure 3:** Quantification of parasite-encoded 18S rRNA in cultures of human hepatocytes infected with *P. falciparum*.

Sporozoites were added at a 2.3:1 sporozoite-to-hepatocyte ratio. Real-time PCR analysis of 18S rRNA expression was done at 48 and 96 hours postinfection. Each culture (n = 3) was analyzed in technical triplicate. (A) 18S copy number normalized using $C_T$ values from GAPDH (B) 18S copy number per culture. (C) Growth curve of uninfected hepatocyte cell lines (n = 3 per time point). Mean +/- SD shown for all graphs.

https://doi.org/10.1371/journal.pone.0129623.g003

## *P. falciparum*-infected HC-04 Hepatocytes can be Specifically Isolated

Infected hepatocytes from mice have been sorted *ex vivo* using *P. yoelii* parasites expressing GFP [40]. We have previously demonstrated that HC-04 cells infected with *P. berghei* ANKA-GFP can be isolated by flow cytometry-based cell sorting based on GFP expression *in vitro*[33]. Utilizing a similar approach we performed infection of HC-04 with *P. falciparum* 3D7HT-GFP sporozoites and sought to verify that $GFP^+$ HC-04 cells contain developing parasites. $GFP^+PI^-$ hepatocytes were sorted and cytospun, followed by immunofluorescence analysis of $GFP^+$ cell populations. Though these experimental procedures may alter native host and parasite morphology it was necessary to perform due to a short doubling time (Fig 3C) of uninfected cells leading to extensive three-dimensional growth of HC-04 cultures. Moreover, sorting and visualization of parasites was needed to validate flow cytometry method as an approach for detection and isolation of *P. falciparum* infected cells.

As shown in Fig 4, $GFP^+$ cells isolated at 96 hours pi contained *P. falciparum* parasites. We used three fluorescent markers to identify the presence of EEFs: parasite-encoded GFP, DAPI to stain both host and parasite DNA, and a monoclonal antibody against *Pf*Hsp70. We observed three patterns in sorted $GFP^+$ cells (Fig 4): (a, b) a GFP-positive, Hsp70 positive parasite located perinuclear and morphologically similar to parasites in primary hepatocytes [12, 13] (c, d) a GFP-positive Hsp70-positive parasite adjacent to a dividing or fragmented host nucleus and (e, f) a GFP-positive Hsp70-positive cell with indistinguishable host DNA. We speculate that the latter group of cells was either devoid of a host nucleus or had a host nucleus containing the parasite. We cannot exclude, however, that some of the hepatocyte cell nuclei were either masked by the parasite or damaged during the cytospin procedure. Since there are no markers available to distinguish each pattern of the parasite by flow cytometry directly, new techniques, such as flow cytometry combined with imaging may become instrumental for further characterization of the parasite patterns in HC-04 cells infected with *P. falciparum*. Punctate DNA staining within the parasites is indicative of development. We do not exclusively observe *P. falciparum* localized in the nucleus of HC-04 cells as has been previously reported for HepG2 [29], however, we cannot formally exclude that this may occur in a subset of $GFP^+$ HC-04 cells.

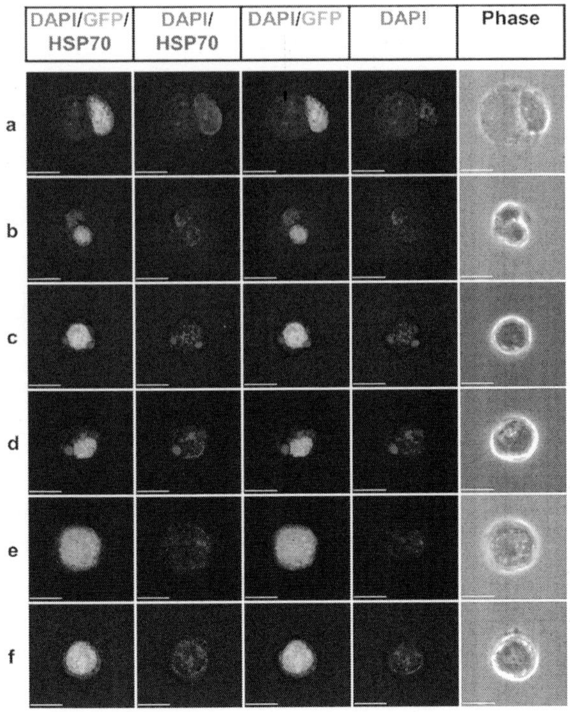


**Figure 4:** Visualization of parasites in HC-04 cells infected with *P. falciparum*.

GFP-positive PI-negative events were isolated by flow cytometry-based cell sorting 96 hours postinfection and cytospun. Representative micrographs of EEFs show DNA staining with DAPI, GFP expression and immunofluorescent detection of *Pf*Hsp70 by anti-Hsp70 monoclonal antibody (Blue = DAPI, Green = GFP, Red = *Pf*Hsp70. Scale bars indicate 10 μm).

https://doi.org/10.1371/journal.pone.0129623.g004

## Primary Human Hepatocytes Support *P. falciparum* 3D7HT-GFP EEFs *in Vitro*

We next asked if *P. falciparum* infection of HC-04 is comparable to that of primary human hepatocytes. Infected hepatocyte cultures from three donors had viable infected ($GFP^+PI^-$) events detectable by flow cytometry at 96 hours pi (Fig 5A). Higher levels of autofluorescence observed in primary hepatocyte cultures required adjustments of flow cytometer voltages prior to acquisition of primary versus HC-04 cells. Consequently, direct comparisons of GFP mean fluorescence intensity between HC-04 and primary hepatocyte cultures are not feasible. Unlike HC-04, GFP+ events detected in primary hepatocyte cultures were within the range of background autofluorescence in the FL1 fluorescence channel (530/30) (Fig 5A). This highlights the necessity of using a ratio of closely related fluorescent emission channels to distinguish GFP+ events from the autofluorescence of primary cells. The relative frequencies of GFP+ events detected in primary hepatocytes at 96 hours pi were 20-fold higher (Fig 5A and 5B) than HC-04. However, the numbers of GFP+PI- cells in each individual culture (Fig 5C) did not significantly differ between HC-04 and two of the primary hepatocyte cultures (donors 4055A and 4059). Infected hepatocyte cultures from donor 4051 supported 3-fold more GFP+cells than HC-04 and primary cells from the two other donors. The drastic difference in percentages of infected cells found in HC-04 versus primary hepatocyte cultures at 96 hours pi (Fig 5A and 5B) was due to extensive proliferation of uninfected HC-04 cells (Fig 3C). To confirm that GFP+ primary hepatocytes contained parasites we sorted and stained cells with an Hsp70-specific antibody (Fig 5D).

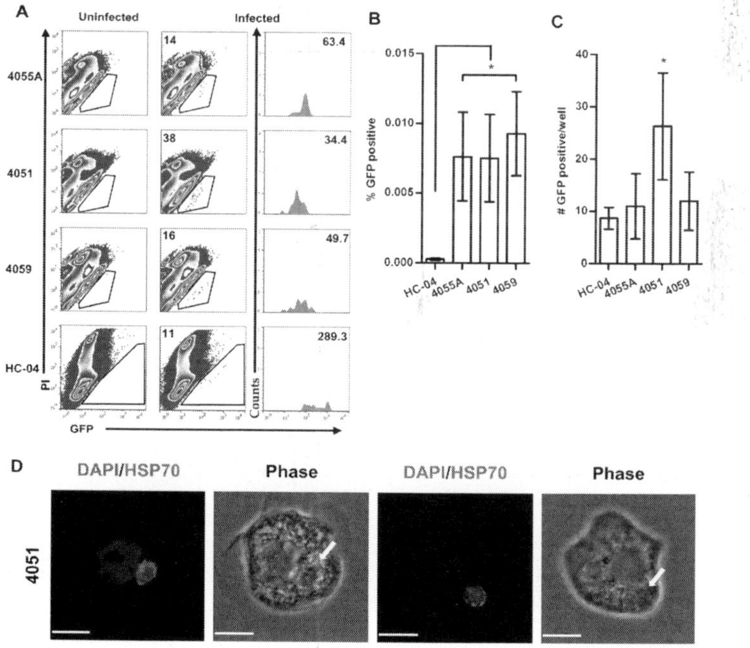


**Figure 5:** Comparison of detection of *P. falciparum* **infection in primary human hepatocytes** and HC-04.

Sporozoites were added at a 1:1 sporozoite-to-hepatocyte ratio. Infected cells were identified by flow cytometry at 96 hours postinfection as GFP-positive events in PI-negative (viable) cell populations (n = 3). Uninfected cultures were used to define the positive gates. (A) Representative gating for detection of GFP events using three human hepatocyte donors and HC-04. Number of GFP-positive PI-negative events indicated (middle panels). Geometric mean of GFP-positive PI-negative events indicated (right panels). (B) Comparisons of the percentage of GFP-positive events and (C) total number of GFP events obtained per well. Mean +/- SD shown. (D) Representative examples of GFP-positive PI-negative events isolated by flow cytometry-based cell sorting 96 hours postinfection and cytospun (Scale bars indicate 10 μm).

https://doi.org/10.1371/journal.pone.0129623.g005

The lack of proliferation of primary hepatocytes allowed for visualization of unsorted EEFs maintained on coverslips throughout infection (Fig 6A). We defined parasites by Hsp70 staining (red) and observed no significant differences in parasite size between hepatocytes from different donors at 96 hours pi (Fig 6B and 6C). Similar to previous studies [12, 13], *Pf*3D7-HTGFP parasites were either adjacent to or touching the host nucleus (Fig 6D). A distance of >10μm was observed between some parasites and the closest host nucleus (Fig 6E).

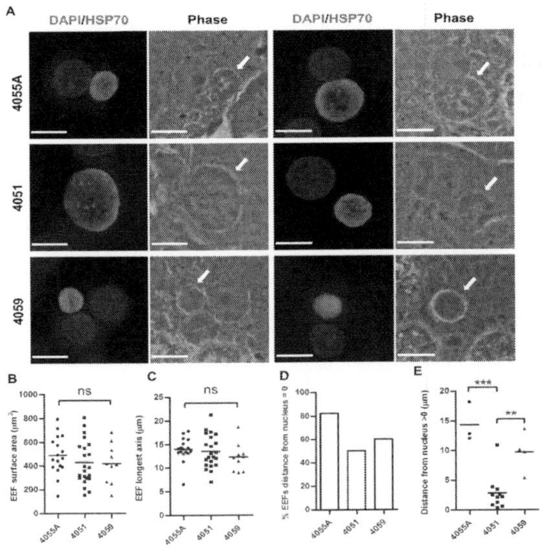


**Figure 6:** Parasite morphology and size in cryopreserved human hepatocytes from three donors.

Sporozoites were added at a 1:1 sporozoite-to-hepatocyte ratio. Cultures were fixed and stained directly on coverslips 96 hours postinfection. (A) Representative micrographs of EEFs observed from each donor. DAPI (blue) represents host and *Plasmodium* nuclei and *Pf*Hsp70 (red) denotes EEF. Arrows indicate EEF in phase contrast micrographs (Scale bar = 10 μM). EEF size was measured by (B) surface area and (C) maximum diameter. (D) Proportion of EEFs observed to be directly adjacent to host nucleus. (E) Of EEFs distant from host nucleus, the distance to the closest nucleus is indicated. Mean shown (horizontal line) for individual dot plots.

https://doi.org/10.1371/journal.pone.0129623.g006

To further substantiate that $GFP^+$ cell populations identified in hepatocyte cultures infected with 3D7HT-GFP are specific to infection with sporozoites, we demonstrated that the percentages of $GFP^+$ cells in both HC-04 and primary hepatocyte cultures significantly drop if sporozoites are pre-incubated with cytochalasin D prior to infection. Also, a clear dependence of the $GFP^+$ cell population size on the sporozoite-to-hepatocyte ratio was observed. These data are in line with our ability demonstrate that sorted $GFP^+$ cell populations are composed of cells containing parasites (Figs 4 and 5).

## Dynamics of *Pf*EEF Detection by Flow Cytometry Differ in HC-04 and Primary Hepatocytes

Our prior experiments were performed at 96 pi to allow for detection of developing parasites. However, the overall ability to detect $GFP^+$ parasites by flow cytometry is a function of their brightness over background and attrition over time. To determine the progression of *Pf*EEF development as well as the sensitivity of flow cytometry-based detection of *Pf*3D7HT-GFP EEFs at different time points post infection, we monitored two parameters: the number of viable infected ($GFP^+PI^-$) cells and the intensity of GFP fluorescence in these cells as a marker of parasite development (Fig 7).

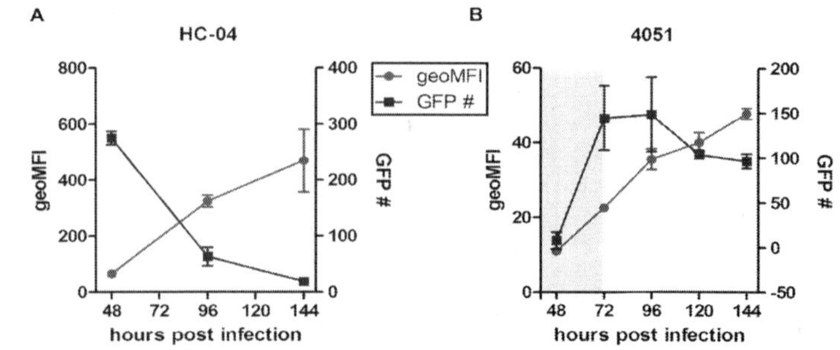


**Figure 7:** Parasite detection and persistence over time.

Sporozoites were added at 0.5:1 sporozoite-to-hepatocyte ratio. Cultures were infected and collected for flow cytometry starting at 48 hours postinfection in duplicate. Plots are shown for the number of GFP-positive events acquired at each time point and the geometric MFI of these events for (A) HC-04 and (B) primary donor 4051 (shading indicates time when a portion of the GFP population is obscured). Mean +/- SD shown.

https://doi.org/10.1371/journal.pone.0129623.g007

The maximal numbers of parasites developing in HC-04 cells were detectable starting at 48 hours pi followed by a steady decrease in the number of GFP$^+$ events per culture over time (Fig 7A). In contrast, at 48 hours pi detection of parasites in primary hepatocytes was partially obscured by background autofluorescence, even when using the ratio of closely related fluorescence channels. Beginning at 72 hours pi a higher number of parasite-infected cells were detectable in primary hepatocyte cultures by flow cytometry (Fig 7B). The relative parasite survival rate between 96–144 hours pi was 29% for HC-04 and 65% in primary hepatocytes (Fig 7). The parasite-specific GFP fluorescence intensity increased steadily in both HC-04 (Fig 7A) and primary hepatocytes (Fig 7B) and reached its maximum at 144 hours pi, suggesting development of the remaining persisting parasites. Based on these data, for all subsequent experiments we chose to monitor *Pf*3D7HT-GFP development at 48 hours pi in HC-04 and at 96 hours pi in primary hepatocytes, when the maximal numbers of parasite-infected cells were observed in their respective cell cultures.

To determine potential applications of our experimental system, we focused on two major questions relevant to *P. falciparum* liver stages: (i) the role of putative host receptors required for efficient infection of human hepatocytes by *P. falciparum* parasites and (ii) the role of host factors limiting development of *P. falciparum* EEFs.

## CD81 is required for *P. falciparum* Infection of Primary Human Hepatocytes, but is not essential for the invasion of HC-04

It has been previously shown that functional antibody-based neutralization of CD81 on the surface of primary human hepatocytes abolishes infection by *P. falciparum* sporozoites *in vitro*as measured by the numbers of *P. falciparum* EEFs detected by immunofluorescence [41]. To address the requirement of CD81 using our experimental model, we first verified the expression of CD81 on the surface of hepatocyte cells lines and primary human hepatocytes. Primary hepatocytes from three donors as well as THLE-2 and -3 cells endogenously expressed surface CD81, whereas neither HC-04 nor HepG2 had detectable expression by flow cytometry (Fig 8A). These data were further confirmed by RT-PCR (data not shown), ruling out our inability to detect low CD81 expression by flow cytometry. These observations in combination with the differential ability of hepatocyte cell lines to support development of *P. falciparum* EEFs (see Figs 1–5) reveal no correlation between the expression of CD81 and the susceptibility of human hepatocyte cell lines to *P. falciparum* infection or to traversal *in vitro*. Lack of CD81 did not preclude infection of HC-04, and endogenous expression of CD81 on THLE-2 and -3 was not sufficient to allow for efficient infection. Accordingly, ectopic expression of CD81 on HepG2 cells did not result in more efficient *P. falciparum* infection (Figs 2 and 3). Similarly, transient expression of CD81 in HC-04 cells did not significantly change the numbers and development of EEFs in these cultures (Fig 8C and 8D).

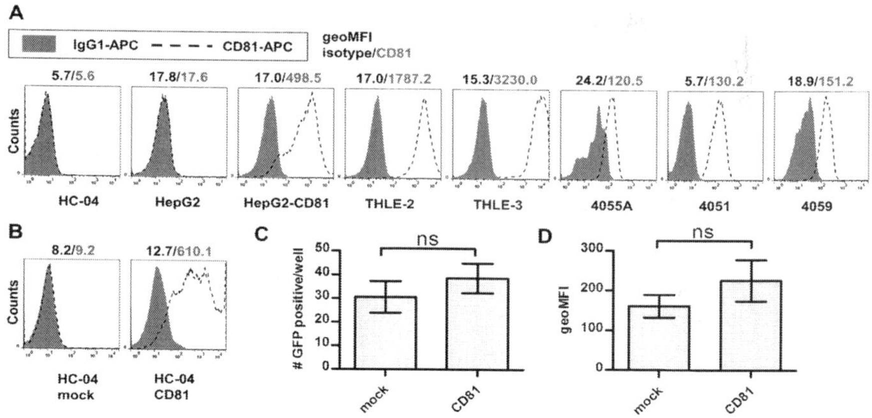


**Figure 8:** Expression of surface CD81 by human hepatocyte cell lines and primary hepatocyte donors.

Surface CD81 was stained using specific antibodies or an isotype control followed by flow cytometry for detection. (A) Representative plots for all cells are shown and geometric MFI is indicated for both isotype (black) and anti-CD81 staining (red). (B) Surface staining of a mock and transient transfection of HC-04. Transiently transfected HC-04 were infected and run on flow cytometry 96 hours postinfection; (C) GFP-positive number and (D) geometric MFI shown. Mean +/- SD shown.

https://doi.org/10.1371/journal.pone.0129623.g008

Using our experimental model, we recapitulated previously described conditions [41] for antibody-based neutralization of CD81 (Fig 9). We found that the CD81-blocking antibody (clone 1D6) had no effect on *P. falciparum* 3D7HT-GFP infection of CD81-negative HC-04, as assessed by percentages and the numbers of infected ($GFP^+PI^-$) cells detected by flow cytometry at 48 hours pi (Fig 9A and 9B). In contrast, the same neutralizing antibody blocked *P. falciparum* infection in primary human hepatocytes (Fig 9C and 9D). Moreover, 1D6 did not alter parasite numbers when added 6 hours pi, suggesting that CD81 is critical only during parasite invasion. These data are in agreement with the original observation by Silvie *et al.*, demonstrating that CD81 blocking significantly reduces invasion of primary human hepatocytes by *P. falciparum* sporozoites [41]. Therefore, though invasion of HC-04 appears to be independent of CD81, we confirm that invasion of primary hepatocytes is CD81-dependent.

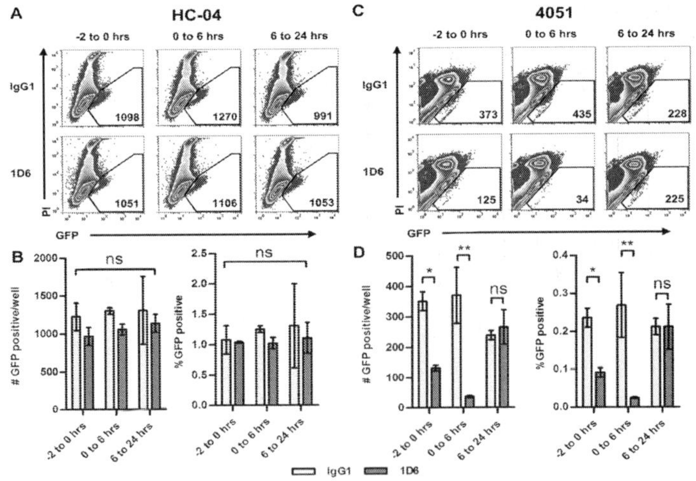


**Figure 9:** Influence of CD81 blocking by mAb 1D6 on *P. falciparum* **hepatocyte infection.**

Sporozoites were added at a 3:1 sporozoite-to-hepatocyte ratio. 1D6 or isotype were added to cultures at 10μg/ml prior to infection (-2 to 0 hours), during invasion (0 to 6 hours) or after invasion (6 to 24 hours). Representative flow plots shown for (A) HC-04 48 hours postinfection and (B) number and percentage of GFP-positive events in duplicate. (C) Flow plots for donor 4051 96 hours postinfection and (D) graphs indicated the number and percentage of GFP-positive events in duplicate. Mean +/- SD shown on all graphs.

https://doi.org/10.1371/journal.pone.0129623.g009

## Humanized CSP-specific Antibody Inhibits Traversal and reduces the Number of *P. falciparum* EEFs in Human Hepatocytes *In Vitro*

The major surface protein of *Plasmodium* sporozoites, the circumsporozoite protein (CSP), is a well-established target for humoral immune responses aimed at neutralization of sporozoite infectivity [42]. To determine the potential utility of our system for the screening of sera for the presence of sporozoite-specific antibodies we characterized the potency of a recently described humanized CSP-specific monoclonal antibody h2A10, produced in a vectored immunoprophylaxis (VIP)-vector transduced mouse [39] (Fig 10). We found that serial dilutions of mouse serum initially containing 1.93 mg/ml of human IgG inhibited parasite traversal in an antibody concentration-dependent manner (Fig 10A). Moreover, h2A10 reduced the relative frequencies and total numbers of $GFP^+PI^-$ cells in both HC-04 and primary hepatocyte cultures (Fig 10B and 10C), although overall inhibition of infection was less prominent in primary cells. Even at a concentration of 3.86 μg/ml (1:500 dilution of h2A10 antibody-containing serum) infection was not completely abolished, suggesting that higher titers of sporozoite antigen-specific antibodies will be required to prevent infection. The intensity of GFP fluorescence in hepatocyte cultures infected along with h2A10 was not significantly different from cultures infected with control sera (Fig 10B and 10C). This suggests that parasites not inhibited by h2A10 during invasion develop normally in infected hepatocytes.

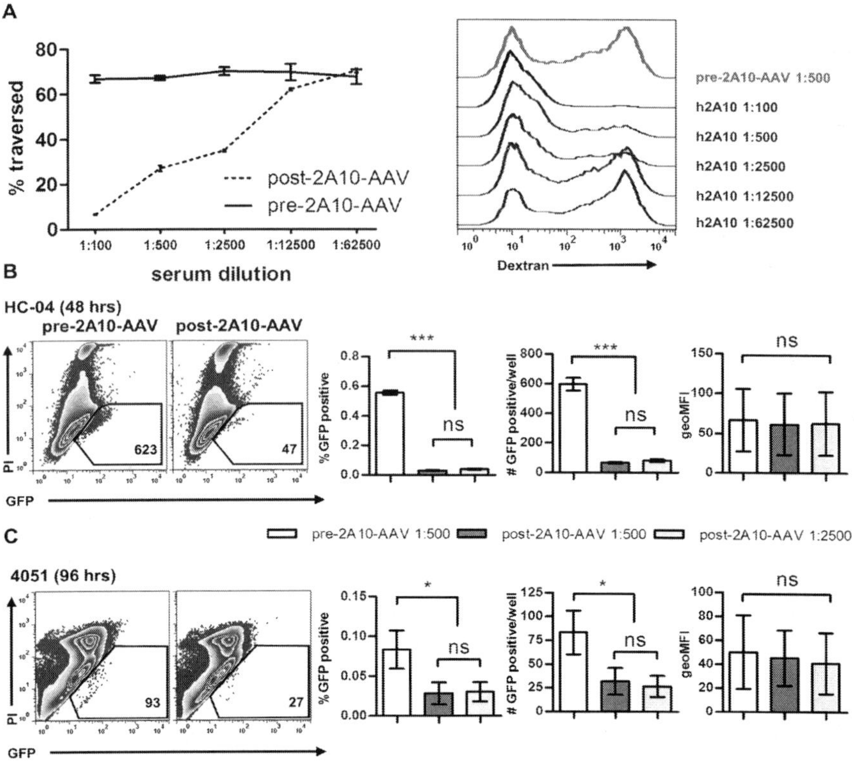


**Figure 10:** Influence of a humanized anti-CSP mAb 2A10 on *P. falciparum* **traversal and invasion.**

Sporozoites were added at a 4.5:1 sporozoite-to-hepatocyte ratio. Cells were infected in the presence of h2A10 or pre-2A10-AAV serum at varying dilutions. (A) Percentage of traversed cells measured by dextran uptake 6 hours postinfection. Representative flow plots for (B) HC-04 48 hours postinfection and (C) donor 4051 96 hours postinfection. Graphs indicated the number, percentage and geometric MFI of GFP-positive events in duplicate. Mean +/- SD shown on all graphs.

https://doi.org/10.1371/journal.pone.0129623.g010

# DISCUSSION

The liver stage of *Plasmodium falciparum* infection, a leading target for generation of protective vaccines, still remains the least characterized part of the parasite lifecycle. Since the demonstration of *P. falciparum*

EEF development in primary human hepatocytes, a number of alternative methodologies relying on microscopy have been identified to study *P. falciparum*liver stages *in vitro* [9, 13]. The generation of fluorescent rodent parasite strains along with a wide range of host cell tropisms has allowed for routine identification and sorting of *Plasmodium*EEFs [40], though such an approach has not been described for *P. falciparum*. Using a recently developed strain of *P. falciparum* expressing GFP [32] we show that live *P. falciparum*-infected hepatocytes are identifiable by flow cytometry and, therefore, can be quantitatively assessed and isolated. This approach allows for sensitive and rapid screening of hepatocyte cell lines and primary human hepatocytes from different donors for their ability to support *Pf*EEF development *in vitro*. Finally, we propose that this approach can be used to quantify the effects of antibodies, cytokines or drugs on parasite invasion and development *in vitro*.

Prior to EEF formation in the liver, *Plasmodium* sporozoites traverse through several host cell types including hepatocytes [23, 26, 43, 44]. We found that *in vitro*, the amount of cells retaining dextran after parasite traversal differed depending on the hepatocyte cell line used (Fig 1). No correlation was observed between the numbers of traversed cells detected 6 hours pi and the frequencies of infected cells detected 96 hours pi (Fig 2). These data suggest that in the cell lines tested the process of invasion and development is distinct from traversal. CSP is shed from the sporozoite surface and is found in both cells permissive or non-permissive to EEF formation [45]. We recently found that flow cytometry-based detection of CSP in *P. falciparum*-infected hepatocyte cultures does not represent an accurate measure of hepatocyte infection, but rather reflects the retention of CSP shed during parasite traversal (Trop S. *et al.*, submitted). Therefore, we used parasite-encoded GFP as a marker for specific identification of developing parasites by flow cytometry. Using a ratio of two fluorescent channels, a FL1 (530/30) for detection of parasite GFP and FL2 (585/42) for detection of propidium iodide (PI) positive cells, we compared five cell lines (Fig 2) and three primary hepatocyte cultures (Fig 5) for their ability to support *Pf*EEF formation. This approach allows discrimination between the dead/PI-positive (FL1<FL2), PI-negative uninfected/GFP-negative (FL1 = FL2) and GFP-positive (FL1>FL2) cells in infected hepatocyte cultures. This tactic was previously used by others [46] and allows for the exclusion of autofluorescent "false positive" events while detecting weak GFP signals that otherwise would be considered "false negative" if only a single parameter was used.

Among all cell lines tested, HC-04 [28] supported the highest absolute number of *Pf*EEFs at 96 hours pi (Fig 2A and 2C). Surprisingly the HepG2 cell line, previously described as being non-permissive for *P. falciparum* sporozoite infection [9], had GFP⁺ events detectable in the viable hepatocyte gate (Fig 2A). The SV40-immortalized hepatocyte lines THLE-2 and THLE-3 [30] had previously not been tested for their ability to support *P. falciparum* infection. Though these lines had a low proliferation rate, making them suitable to study the late stages of *Pf*EEFs, both THLE-2 and THLE-3 showed a minimal ability to support *Pf* liver stages in *vitro* (Fig 2A and 2C). In addition to the numbers of GFP⁺ events detectable in infected hepatocyte cultures, we compared the fluorescent intensity of these events as a measure of parasite development in different cell lines. The geometric mean fluorescence intensity (geoMFI) values were assessed in gated GFP⁺ cell populations detected in infected cultures and derived from each well separately. To further verify the presence of developing *P. falciparum* EEFs we used quantitative real-time PCR for detection of *Pf*18S at two time points postinfection (Fig 3). Of the five cell lines tested, a time-dependent increase in *Pf*18S gene copy number per each individual well was detected only in HC-04 cultures (Fig 3B), indicating that *P. falciparum* parasites can develop in these cells.

Primary human hepatocytes remain the gold standard for generation of *Pf*EEFs *in vitro*. However, recent work has questioned the susceptibility of all primary hepatocytes to *P. falciparum* infection, at least *in vitro* [13]. Using flow cytometry we demonstrated, that cryopreserved, commercially available human hepatocytes obtained from 3 different donors were permissive for infection with *P. falciparum* 3D7HT-GFP sporozoites. Though efficient detection of parasite-specific GFP signal at the early stages of infection was hampered due to high autofluorescence in normal hepatocytes [9, 47, 48], primary cells infected with *Pf*3D7HT-GFP were reproducibly detectable by flow cytometry between 72–144 hours pi (Fig 7B). It needs to be seen if flow cytometry based detection can be applied to the *ex vivo* detection of the liver stage parasites in the recently developed humanized mouse models of *P. falciparum*infection [16].

To address the dynamics of 3D7HT-GFP parasite detection and development in hepatocytes *in vitro* we used flow cytometry to monitor changes in parasite numbers and parasite-specific GFP fluorescence (MFI) as a reflection of parasite expansion within infected hepatocytes. The maximum numbers of GFP⁺ events in HC-04 cultures were detected at 48 hours pi followed by a rapid decline (Fig 7A). In contrast, a larger proportion of infected cells persisted from 96 to 144 hours pi in primary hepatocytes (Fig 7B). A similar trend has been observed by others [13]. Importantly, an increase in GFP fluorescence over time was seen in infected cells that persisted in both HC-04 and primary hepatocyte cultures, indicative of parasite development. Taken together, these results show that *P. falciparum* parasites can develop in both HC-04 and primary hepatocytes.

To visualize the presence of parasites in $GFP^+$ cells and to demonstrate the ability to selectively isolate live parasites from bulk infected cultures we sorted $GFP^+$ HC-04 cells and primary hepatocytes (Figs 4 and 5). It does not appear that *Pf*EEFs developing in HC-04 cells exclusively localize to the host nucleus, a phenomenon that has been previously observed in HepG2 cells [29]. Using GFP and *Pf*Hsp70 as parasite-specific markers, we identified three patterns of *P. falciparum* parasites in HC-04 cells at 96 hours pi. Parasites were found either adjacent to an intact host nucleus, surrounded by a fragmenting or dividing host nuclei or without an apparent host nucleus distinct from the parasite DNA (Fig 4). Additional studies are needed to describe phenotypes of parasites detected in HC-04 cells at the later time points of infection. In contrast to parasites developing in HC-04, *Pf*EEFs detected in sorted $GFP^+$primary hepatocytes were mainly adjacent to the intact host nucleus easily distinguishable from the parasite nuclei (Fig 5D).

In contrast to proliferating HC-04 cells that rapidly form three-dimensional cell cultures composed of uninfected cells overlaying parasite-infected cells, non-proliferating primary hepatocytes permit identification of EEFs by fluorescent microscopy without prior sorting (Fig 6). Unlike sorting and cytospin procedures, direct fixation of infected cells on coverslips better preserves both parasite and host hepatocyte morphology. In all three primary infected hepatocyte cultures, *Pf*Hsp70-specific antibody identified developing parasites (Fig 6), but GFP fluorescence was not detectable in these cells by fluorescence microscopy (data not shown). As discussed above, the inability of GFP signal to be detected by fluorescence microscopy is likely due to high autofluorescence in normal hepatocytes [9]. At 96 hours pi *Pf*EEFs developing in primary hepatocyte cultures were mainly adjacent to the host nucleus and comparable in size to parasites described *in vivo* [16] (Fig 6).

After establishing a model of *in vitro Pf* EEF detection and verifying the presence of parasites in both primary hepatocyte and HC-04 cell cultures, we designed several experiments to explore potential applications of our model for both basic and translational studies. Ligands on the human hepatocyte cell surface facilitating invasion by *P. falciparum* sporozoites are still poorly understood. We re-examined the necessity of CD81 expression for invasion by *P. falciparum*sporozoites. Previous studies determined that blocking of CD81 during *P. falciparum* infection reduces EEF numbers in primary hepatocytes [41], however expression of CD81 on non-permissive cell types was insufficient to confer invasion [29]. Here we show that endogenous surface expression of CD81 (Fig 8A) did not correlate with *P. falciparum* EEF numbers in the five cell lines tested (Fig 2). Primary hepatocytes derived from 3 human donors were permissive for *P. falciparum* infection and had expression of CD81 on the cell surface (Fig 8A). HC-04, the most permissive to *P. falciparum* infection among all cell lines tested, lacks CD81 and transient ectopic expression of CD81 did not alter susceptibility of HC-04 cells to invasion (Fig 8C and 8D). Taking advantage of the ability to quantitatively assess EEF development by flow cytometry, we re-examined the requirement for CD81 expression during *P. falciparum*sporozoite invasion (Fig 9). We found that the CD81-blocking monoclonal antibody 1D6 did not alter *P. falciparum* infection of HC-04 cells (Fig 9A and 9B). However, in agreement with the original finding by Silvie *et al.* [41], the same antibody decreased the *Pf*EEF numbers in primary human hepatocyte cultures when used during sporozoite invasion (Fig 9C and 9D). The inability of 1D6 to alter sporozoite invasion of CD81-negative HC-04 suggests that 1D6 has no unspecific effect on sporozoites. In agreement, pre-incubation of primary hepatocytes with 1D6 followed by its removal prior to infection blocked sporozoite invasion. Additionally, no effect of 1D6 was observed when added six hours after infection in either HC-04 or primary hepatocyte cultures. Our findings suggest that different host cell surface receptors and/or distinct parasite invasion pathways can be utilized by *P. falciparum* sporozoites to establish liver stage infection. One possibility is that differences in heparan sulfate proteoglycans (HSPG) between HC-04 and primary hepatocytes mediate a differential requirement for CD81. It has been established that CSP binds to HSPGs and influences sporozoite invasion [49–51] but does so more prominently under dynamic flow conditions [52]. Alternatively, the inability of CD81 antibodies to alter $GFP^+$parasite numbers detected in infected HC-04 cell cultures by flow cytometry could reflect the persistence of transmigrating sporozoites trapped in a proportion of hepatocytes. Novel technologies based on combining flow cytometry and fluorescence microscopy may prove useful in addressing this issue. We propose that our experimental approach can facilitate the discovery of additional human host factors or further characterize the importance of

HSPGs during *P. falciparum* sporozoite invasion.

Next we demonstrated the utility of these methods to quantify the effects of anti-sporozoite antibodies on traversal and invasion. Sterilizing immunity to the liver stage of *Plasmodium*infection does not occur during natural infection [53]. It has been well established that antibodies directed against the sporozoite are capable of blocking parasite motility *in vitro* and *in vivo* and can consequently reduce parasite invasion into hepatocytes [54–56]. The RTS,S vaccine is capable of inducing anti-CSP antibodies able to protect against *P. falciparum*infection in a dose dependent manner [17]. Inhibition of parasite motility as measured by traversal *in vitro* has been used to characterize humoral responses against the sporozoite [57]. However, to establish a correlation between the effects exerted by parasite-specific antibodies on traversal with effects on formation of *Pf*EEFs has required a humanized mouse model [17, 57]. Here, using serum from a vectored immunoprophylaxis (VIP)-vector transduced mouse producing a humanized *Pf*CSP-specific monoclonal antibody [39], we demonstrate an inhibition of both traversal and infection by *P. falciparum* sporozoites *in vitro* (Fig 10). Since our approach is not limited to a specific parasite antigen, it may be applied to screening of sera from human cohorts immunized with radiation attenuated (RAS), genetically attenuated (GAS) sporozoites, or sporozoites combined with chloroquine chemoprophylaxis (CPS) based protection [20, 58]. *In vitro* methods enabling objective evaluation of anti-sporozoite humoral immune responses are particularly advantageous because they require considerably less material than humanized mouse models and can include multiple biological replicates. However, it should be stated that *in vitro* models of infection lack the complex hepatic architecture and cannot entirely recapitulate the natural infection.

Previously, detection of *Pf*EEFs by flow cytometry has only been possible using rodent *Plasmodium* species [9, 24, 33, 40]. Although the use of rodent parasites and mouse models of *Plasmodium* infection will continue to be a cornerstone of malaria research, advancements in methods for studying *Pf*EEFs *in vitro* are required to strengthen our understanding of basic parasite biology and to provide objective and quantifiable protocols to inform translational research. HC-04 and primary hepatocytes constitute complementary *in vitro* models to study *Pf*EEF formation. Among five cell lines tested, HC-04 is able to support the greatest number of developing parasites. However, uninfected cells in HC-04 cultures rapidly proliferate, complicating parasite detection after 72–96 hours pi. On the other hand, parasite traversal can be quantified in these cells early after infection and fluorescent parasites are clearly detectable by flow cytometry starting at 48 hours pi. Consequently, HC-04 is a suitable host cell to screen for the efficacy of interventions aimed at blocking the motility of sporozoites, such as anti-parasite antibodies, which can be assessed in traversal assays. However, the HC-04 cell line is not fully suitable for study of late *Pf* liver stages.

In contrast to HC-04 cells, primary hepatocytes do not proliferate *in vitro*, however, the *Pf*EEFs can be clearly detected in these cells by flow cytometry only beginning at 72 hours pi. Though primary hepatocytes will continue to be the standard for study of *Pf*EEFs, *de novo* generation or identification of additional human hepatocyte cell lines with low doubling times and limited autofluorescence still remains a priority. The development of a *P. falciparum* line expressing GFP through the parasite's lifecycle [32] was paramount in our ability to reproducibly detect and quantify *Pf*EEFs by flow cytometry. The generation of a novel *P. falciparum* strain expressing a fluorophore brighter than GFP and/or use of a stronger liver stage promoter will facilitate flow cytometry-based detection and isolation of *P. falciparum* parasites in human hepatocytes even before 48 hours postinfection.

## ACKNOWLEDGMENTS

We thank the Johns Hopkins Malaria Research Institute (JHMRI) Insectary and Parasite Core facilities for mosquito rearing and gametocyte cultures. We thank Dr. Godfree Mlambo and Dr. Abhai Tripathi (JHMRI) for technical support. Dr. Gary Ketner and Dr. Cailin Deal (JHMRI) kindly provided mouse serum containing h2A10 antibody.

## AUTHOR CONTRIBUTIONS

Conceived and designed the experiments: PD JL. Performed the experiments: PD JL ST JM HZ. Analyzed the data: PD JL ST JM. Contributed reagents/materials/analysis tools: HZ MS. Wrote the paper: PD JL.

## REFERENCES

1. World Health Organization. World Malaria Report 2013 Geneva, Switzerland 2013. Available: http://www. who. int/malaria/publications/world_malaria_report_2013/report/en/.
2. Vaughan AM, Aly AS, Kappe SH. Malaria parasite pre-erythrocytic stage infection: gliding and hiding. Cell Host Microbe. 2008;4(3):209–18. pmid:18779047.
3. Boddey JA, Cowman AF. *Plasmodium* nesting: remaking the erythrocyte from the inside out. Annu Rev Microbiol. 2013;67:243–69. pmid:23808341.
4. Trager W, Jensen JB. Human malaria parasites in continuous culture. Science. 1976;193(4254):673–5 pmid:781840.
5. Costa FT, Avril M, Nogueira PA, Gysin J. Cytoadhesion of *Plasmodium falciparum*-infected erythrocytes and the infected placenta: a two-way pathway. Braz J Med Biol Res. 2006;39(12):1525–36 pmid:17160261.
6. Baum J. A complete molecular understanding of malaria parasite invasion of the human erythrocyte: are we there yet? Pathog Glob Health. 2013;107(3):107. pmid:23683362.
7. Bei AK, Duraisingh MT. Functional analysis of erythrocyte determinants of *Plasmodium*infection. Int J Parasitol. 2012;42(6):575–82. pmid:22726752.
8. Duffy PE, Sahu T, Akue A, Milman N, Anderson C. Pre-erythrocytic malaria vaccines: identifying the targets. Expert Rev Vaccines. 2012;11(10):1261–80. pmid:23176657.
9. Prudêncio M, Mota MM, Mendes AM. A toolbox to study liver stage malaria. Trends Parasitol. 2011;27(12):565–74. pmid:22015112.
10. Hollingdale MR, Leland P, Schwartz AL. *In vitro* cultivation of the exoerythrocytic stage of *Plasmodium berghei* in a hepatoma cell line. Am J Trop Med Hyg. 1983;32(4):682–4 pmid:6349397.
11. Mota MM, Rodriguez A. *Plasmodium yoelii*: efficient *in vitro* invasion and complete development of sporozoites in mouse hepatic cell lines. Exp Parasitol. 2000;96(4):257–9. pmid:11162379.
12. Mazier D, Beaudoin RL, Mellouk S, Druilhe P, Texier B, Trosper J, et al. Complete development of hepatic stages of *Plasmodium falciparum in vitro*. Science. 1985;227(4685):440–2 pmid:3880923.
13. March S, Ng S, Velmurugan S, Galstian A, Shan J, Logan DJ, et al.

A microscale human liver platform that supports the hepatic stages of *Plasmodium falciparum* and *vivax*. Cell Host Microbe. 2013;14(1):104–15. pmid:23870318.

14. Sacci JB Jr. , Schriefer ME, Resau JH, Wirtz RA, Detolla LJ Jr. , Markham RB, et al. Mouse model for exoerythrocytic stages of *Plasmodium falciparum* malaria parasite. Proc Natl Acad Sci U S A. 1992;89(9):3701–5 pmid:1570289.
15. Sacci JB Jr. , Alam U, Douglas D, Lewis J, Tyrrell DL, Azad AF, et al. *Plasmodium falciparum* infection and exoerythrocytic development in mice with chimeric human livers. Int J Parasitol. 2006;36(3):353–60. pmid:16442544.
16. Vaughan AM, Mikolajczak SA, Wilson EM, Grompe M, Kaushansky A, Camargo N, et al. Complete *Plasmodium falciparum* liver-stage development in liver-chimeric mice. J Clin Invest. 2012;122(10):3618–28. pmid:22996664.
17. Foquet L, Hermsen CC, van Gemert GJ, Van Braeckel E, Weening KE, Sauerwein R, et al. Vaccine-induced monoclonal antibodies targeting circumsporozoite protein prevent *Plasmodium falciparum* infection. J Clin Invest. 2014;124(1):140–4. pmid:24292709.
18. Hafalla JC, Silvie O, Matuschewski K. Cell biology and immunology of malaria. Immunol Rev. 2011;240(1):297–316. pmid:21349101.
19. Spence PJ, Langhorne J. T cell control of malaria pathogenesis. Curr Opin Immunol. 2012;24(4):444–8. pmid:22658628.
20. Roestenberg M, McCall M, Hopman J, Wiersma J, Luty AJ, van Gemert GJ, et al. Protection against a malaria challenge by sporozoite inoculation. N Engl J Med. 2009;361(5):468–77. pmid:19641203.
21. Schwenk RJ, Richie TL. Protective immunity to pre-erythrocytic stage malaria. Trends Parasitol. 2011;27(7):306–14. pmid:21435951.
22. Tse SW, Radtke AJ, Zavala F. Induction and maintenance of protective CD8+ T cells against malaria liver stages: implications for vaccine development. Mem Inst Oswaldo Cruz. 2011;106 Suppl 1:172–8 pmid:21881772.
23. Mota MM, Pradel G, Vanderberg JP, Hafalla JC, Frevert U, Nussenzweig RS, et al. Migration of *Plasmodium* sporozoites through cells before infection. Science. 2001;291(5501):141–4. pmid:11141568.
24. Prudêncio M, Rodrigues CD, Ataide R, Mota MM. Dissecting *in vitro* host cell infection by *Plasmodium* sporozoites using flow cytometry.

Cell Microbiol. 2008;10(1):218–24. pmid:17697130.

25. Frevert U, Engelmann S, Zougbede S, Stange J, Ng B, Matuschewski K, et al. Intravital observation of *Plasmodium berghei* sporozoite infection of the liver. PLoS Biol. 2005;3(6):e192. pmid:15901208.
26. Amino R, Giovannini D, Thiberge S, Gueirard P, Boisson B, Dubremetz JF, et al. Host cell traversal is important for progression of the malaria parasite through the dermis to the liver. Cell Host Microbe. 2008;3(2):88–96. pmid:18312843.
27. Sidjanski S, Vanderberg JP. Delayed migration of *Plasmodium* sporozoites from the mosquito bite site to the blood. Am J Trop Med Hyg. 1997;57(4):426–9 pmid:9347958.
28. Sattabongkot J, Yimamnuaychoke N, Leelaudomlipi S, Rasameesoraj M, Jenwithisuk R, Coleman RE, et al. Establishment of a human hepatocyte line that supports *in vitro*development of the exo-erythrocytic stages of the malaria parasites *Plasmodium falciparum* and *P. vivax*. Am J Trop Med Hyg. 2006;74(5):708–15 pmid:16687667.
29. Silvie O, Greco C, Franetich JF, Dubart-Kupperschmitt A, Hannoun L, van Gemert GJ, et al. Expression of human CD81 differently affects host cell susceptibility to malaria sporozoites depending on the *Plasmodium* species. Cell Microbiol. 2006;8(7):1134–46. pmid:16819966.
30. Pfeifer AM, Cole KE, Smoot DT, Weston A, Groopman JD, Shields PG, et al. Simian virus 40 large tumor antigen-immortalized normal human liver epithelial cells express hepatocyte characteristics and metabolize chemical carcinogens. Proc Natl Acad Sci U S A. 1993;90(11):5123–7 pmid:7685115.
31. Young L, Sung J, Stacey G, Masters JR. Detection of Mycoplasma in cell cultures. Nat Protoc. 2010;5(5):929–34. pmid:20431538.
32. Talman AM, Blagborough AM, Sinden RE. A *Plasmodium falciparum* strain expressing GFP throughout the parasite›s life-cycle. PLoS One. 2010;5(2):e9156. pmid:20161781.
33. Ma J, Trop S, Baer S, Rakhmanaliev E, Arany Z, Dumoulin P, et al. Dynamics of the major histocompatibility complex class I processing and presentation pathway in the course of malaria parasite development in human hepatocytes: implications for vaccine development. PLoS One. 2013;8(9):e75321. pmid:24086507.
34. Schoone GJ, Oskam L, Kroon NC, Schallig HD, Omar SA. Detection and quantification of *Plasmodium falciparum* in blood samples

using quantitative nucleic acid sequence-based amplification. J Clin Microbiol. 2000;38(11):4072–5 pmid:11060070.

35. Tsuji M, Mattei D, Nussenzweig RS, Eichinger D, Zavala F. Demonstration of heat-shock protein 70 in the sporozoite stage of malaria parasites. Parasitol Res. 1994;80(1):16–21 pmid:8153120.
36. Nardin EH, Nussenzweig V, Nussenzweig RS, Collins WE, Harinasuta KT, Tapchaisri P, et al. Circumsporozoite proteins of human malaria parasites *Plasmodium falciparum*and *Plasmodium vivax*. J Exp Med. 1982;156(1):20–30 pmid:7045272.
37. Anker R, Zavala F, Pollok BA. VH and VL region structure of antibodies that recognize the (NANP)3 dodecapeptide sequence in the circumsporozoite protein of *Plasmodium falciparum*. Eur J Immunol. 1990;20(12):2757–61. pmid:2125276.
38. Zavala F, Tam JP, Hollingdale MR, Cochrane AH, Quakyi I, Nussenzweig RS, et al. Rationale for development of a synthetic vaccine against *Plasmodium falciparum*malaria. Science. 1985;228(4706):1436–40 pmid:2409595.
39. Deal C, Balazs AB, Espinosa DA, Zavala F, Baltimore D, Ketner G. Vectored antibody gene delivery protects against *Plasmodium falciparum* sporozoite challenge in mice. Proc Natl Acad Sci U S A. 2014;111(34):12528–32. pmid:25114213.
40. Tarun AS, Baer K, Dumpit RF, Gray S, Lejarcegui N, Frevert U, et al. Quantitative isolation and *in vivo* imaging of malaria parasite liver stages. Int J Parasitol. 2006;36(12):1283–93. pmid:16890231.
41. Silvie O, Rubinstein E, Franetich JF, Prenant M, Belnoue E, Renia L, et al. Hepatocyte CD81 is required for *Plasmodium falciparum* and *Plasmodium yoelii* sporozoite infectivity. Nat Med. 2003;9(1):93–6. pmid:12483205.
42. Macete EV, Sacarlal J, Aponte JJ, Leach A, Navia MM, Milman J, et al. Evaluation of two formulations of adjuvanted RTS, S malaria vaccine in children aged 3 to 5 years living in a malaria-endemic region of Mozambique: a Phase I/IIb randomized double-blind bridging trial. Trials. 2007;8:11. pmid:17386091.
43. Tavares J, Formaglio P, Medvinsky A, Ménard R, Amino R. Imaging sporozoite cell traversal in the liver of mice. Methods Mol Biol. 2013;923:401–10. pmid:22990794.

44. Ishino T, Chinzei Y, Yuda M. A *Plasmodium* sporozoite protein with a membrane attack complex domain is required for breaching the liver sinusoidal cell layer prior to hepatocyte infection. Cell Microbiol. 2005;7(2):199–208. pmid:15659064.

45. Hügel FU, Pradel G, Frevert U. Release of malaria circumsporozoite protein into the host cell cytoplasm and interaction with ribosomes. Mol Biochem Parasitol. 1996;81(2):151–70 pmid:8898331.

46. Li F, Yang M, Wang L, Williamson I, Tian F, Qin M, et al. Autofluorescence contributes to false-positive intracellular Foxp3 staining in macrophages: a lesson learned from flow cytometry. J Immunol Methods. 2012;386(1–2):101–7. pmid:23046996.

47. Croce AC, De Simone U, Freitas I, Boncompagni E, Neri D, Cillo U, et al. Human liver autofluorescence: an intrinsic tissue parameter discriminating normal and diseased conditions. Lasers Surg Med. 2010;42(5):371–8. pmid:20583250.

48. Croce AC, Ferrigno A, Vairetti M, Bertone R, Freitas I, Bottiroli G. Autofluorescence properties of isolated rat hepatocytes under different metabolic conditions. Photochem Photobiol Sci. 2004;3(10):920–6. pmid:15480482.

49. Frevert U, Sinnis P, Cerami C, Shreffler W, Takacs B, Nussenzweig V. Malaria circumsporozoite protein binds to heparan sulfate proteoglycans associated with the surface membrane of hepatocytes. J Exp Med. 1993;177(5):1287–98 pmid:8478608.

50. Frevert U, Sinnis P, Esko JD, Nussenzweig V. Cell surface glycosaminoglycans are not obligatory for *Plasmodium berghei* sporozoite invasion *in vitro*. Mol Biochem Parasitol. 1996;76(1–2):257–66 pmid:8920011.

51. Pradel G, Garapaty S, Frevert U. Proteoglycans mediate malaria sporozoite targeting to the liver. Mol Microbiol. 2002;45(3):637–51 pmid:12139612.

52. Pinzon-Ortiz C, Friedman J, Esko J, Sinnis P. The binding of the circumsporozoite protein to cell surface heparan sulfate proteoglycans is required for *Plasmodium*sporozoite attachment to target cells. J Biol Chem. 2001;276(29):26784–91. pmid:11352923.

53. Tran TM, Li S, Doumbo S, Doumtabe D, Huang CY, Dia S, et al. An intensive longitudinal cohort study of Malian children and adults

reveals no evidence of acquired immunity to *Plasmodium falciparum* infection. Clin Infect Dis. 2013;57(1):40–7. pmid:23487390.

54. Vanderberg JP. Studies on the motility of *Plasmodium* sporozoites. J Protozool. 1974;21(4):527–37 pmid:4138523.
55. Stewart MJ, Nawrot RJ, Schulman S, Vanderberg JP. *Plasmodium berghei* sporozoite invasion is blocked *in vitro* by sporozoite-immobilizing antibodies. Infect Immun. 1986;51(3):859–64 pmid:3512436.
56. Vanderberg JP, Frevert U. Intravital microscopy demonstrating antibody-mediated immobilisation of *Plasmodium berghei* sporozoites injected into skin by mosquitoes. Int J Parasitol. 2004;34(9):991–6. pmid:15313126.
57. Behet MC, Foquet L, van Gemert GJ, Bijker EM, Meuleman P, Leroux-Roels G, et al. Sporozoite immunization of human volunteers under chemoprophylaxis induces functional antibodies against pre-erythrocytic stages of *Plasmodium falciparum*. Malar J. 2014;13:136. pmid:24708526.
58. Butler NS, Vaughan AM, Harty JT, Kappe SH. Whole parasite vaccination approaches for prevention of malaria infection. Trends Immunol. 2012;33(5):247–54. pmid:22405559.

# CHAPTER 8

# APPLICATIONS OF FLOW CYTOMETRY TO CHARACTERIZE BACTERIAL PHYSIOLOGICAL RESPONSES

**Verónica Ambriz-Aviña, Jorge A. Contreras-Garduño, and Mario Pedraza-Reyes**

Departamento de Biología, División de Ciencias Naturales y Exactas, Universidad de Guanajuato, Noria Alta S/N, 36050 Guanajuato, GTO, Mexico

## ABSTRACT

Although reports of flow cytometry (FCM) applied to bacterial analysis are increasing, studies of FCM related to human cells still vastly outnumber other reports. However, current advances in FCM combined with a new generation of cellular reporter probes have made this technique suitable for analyzing physiological responses in bacteria. We review how FCM has been applied to characterize distinct physiological conditions in bacteria including

**Citation:** Verónica Ambriz-Aviña, Jorge A. Contreras-Garduño, and Mario Pedraza-Reyes, "Applications of Flow Cytometry to Characterize Bacterial Physiological Responses," BioMed Research International, vol. 2014, Article ID 461941, 14 pages, 2014. https://doi.org/10.1155/2014/461941.

responses to antibiotics and other cytotoxic chemicals and physical factors, pathogen-host interactions, cell differentiation during biofilm formation, and the mechanisms governing development pathways such as sporulation. Since FCM is suitable for performing studies at the single-cell level, we describe how this powerful technique has yielded invaluable information about the heterogeneous distribution of differently and even specialized responding cells and how it may help to provide insights about how cell interaction takes place in complex structures, such as those that prevail in bacterial biofilms.

## INTRODUCTION

The study of bacterial physiological responses using approaches that assess the overall population response considering it as homogeneous is becoming a thing of the past. Current experimental evidence indicates that an ordinary laboratory culture considered as being composed of isogenic bacteria is actually constituted of heterogeneous subpopulations [1, 2] that respond differentially to changes in their environment [3]. Although an invaluable amount of knowledge has been obtained by studying batch cultures, it should be considered that intracell heterogeneity and cell-to-cell interaction (e.g., quorum sensing) exist in an isogenic bacterial culture. Therefore, analysis focused on subpopulations combined with single-cell analysis techniques will help progress our understanding of the bacterial subpopulation behavior in laboratory cultures.

Current available technologies like flow cytometry (FCM) provide more information regarding the individual events that may rule out the overall population response [4]. As reviewed and discussed by Lidstrom and Konopka [5], a normal distribution in the response across the bacterial cells in a culture does not exist, but there are on, off, and intermediate states that depend on a threshold response mechanism, giving rise to physiologically distinct populations. The extent or probability that a well-differentiated subpopulation response could impact on the overall population behavior or fate of a bacterial culture will depend on the nature of the response itself. For instance, when the subpopulation produces and secretes an inhibitor, a growth factor, or an autoinducer into the medium, it causes a response from the rest of the cells [6]. The origins of such heterogeneity are, according to the literature [7], due to differences in microenvironments [8], created by a large number of bacterial cells growing in and modifying a culture, giving rise to what is called extrinsic noise, in combination with intrinsic cellular noise, due to the fact that each cell possess a different spatiotemporal distribution of cellular components, that is, the threshold levels of molecules that switch on/off gene expression arose at different times in each cell, since gene expression is stochastic [9].

In this context, if we are interested in assessing a bacterial response to certain stimuli, using a fluorescent tracer, and we find no difference between the treated and non-treated samples, it may be due to an absence of response or perhaps to a response in a few percentage of the cells. In a distinct scenario, we could measure the treated culture, if we find in the treated sample half the fluorescence compared to the non-treated sample, we may interpret that all the cells diminished the response by half or perhaps that half the cells do not fluoresce at all or perhaps all the intermediary situations. In order to solve this question, it is necessary to measure the fluorescence emitted by each cell in the culture.

FCM is a useful tool for accomplishing this purpose; this technology was first used in the 1960's [10], and since then a huge increase in its development and applications in different fields can be found in the literature. In 1996, Davey and Kell wrote a comprehensive review about the application of FCM in studying heterogeneous microbial populations [10]; therefore, the present review is primarily focused on the information published after that report, emphasizing the applications of FCM when studying bacterial physiological responses.

## APPLICATION OF FLOW CYTOMETRY

### Flow Cytometry: An Overview

A flow cytometer is an apparatus that makes cells or micrometric particles pass through an interrogation point, where a laser beam impacts them and the light that the particle absorbs, scatters or emits due to its intrinsic or extrinsic physical properties are measured. There are a variety of bibliographic resources [11, 12] that explain at length how a flow cytometer works, but for practical purposes, here we only present a brief and simple description. A flow cytometer is composed of three major parts: fluidics, optics, and electronics systems (Figure 1). The goal of the fluidics system is to make the cell or particle of interest pass through the interrogation point, one by one, in the center of a core stream, within which the diameter of the particle's trajectory is controlled by means of the laminar flow exerted by a surrounding sheath stream so that ideally just one cell passes at a time. The optics system consists of a light source: the most typical one is a 488 nm, (blue) Argon ion laser; this beam is focused by means of lens to impact on the cell in the interrogation point. The particle under study scatters light to all the angles, and the light scattered at acute angles, called forward scatter, is (in general) indicative of a particle's size. The light scattered at wide angles (90°), which is called side scatter, is proportional to the particles' roughness and complexity. Forward scattered light is detected in front of the incident laser beam trajectory by a photodiode or by a photomultiplier tube (PMT), and side scattered light and fluorescence are collected, collimated, and directed 90° through a pathway between a series of dichroic mirrors or beam splitters that permit the pass of certain, unwanted wavelength light and reflect the wanted light to a band pass filter (red, orange, yellow, and green light) and finally to a PMT (Figure 1). In the electronics system, the PMT collects photons and expels electrons, amplifying the signal because a PMT may produce a few hundred electrons for each photon. The photo detectors then produce brief current pulses that are amplified, converted to voltage, and finally converted to numbers by the analog signal processing electronics. There are several ways to represent all the hundreds of events generated per second, and ultimately, although the user knows that he puts in cells in a flow cytometer, after the analysis he obtains a series of numbers and plots to interpret [11]. Here is where a subsequent fluorescence activated cell sorting (FACS) and/or complementary (e.g., microscopy) analysis would be useful.

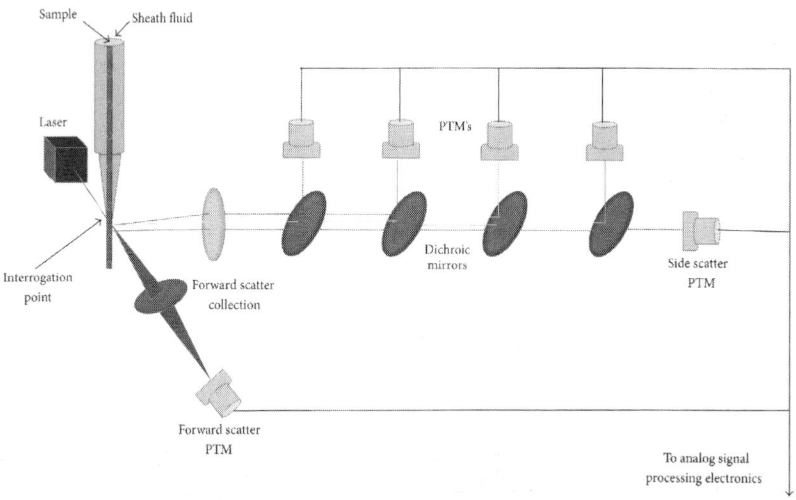


**Figure 1:** Drawing of the main elements that compose a flow cytometer.

As noted above, FCM provides useful information regarding the intrinsic characteristics of a cell or particle; thus, the amount of light scattered at low angles or forward scattering (FS) keeps in most cases a direct relationship with the size of the particle, whereas the complexity, granularity, and protein content can be generally estimated on basis of the light scattered at high angles, a property referred to as side scattering (SS) [11]. Although light scattering is a function of intrinsic characteristics of cells it is difficult to distinguish between different bacterial species employing only this property, therefore, to accomplish this task, cells must be marked with fluorescent dyes [13]. However, light scattering has been successfully used to determine distinct biological properties in bacteria, including, poli(β-hydroxybutyrate) accumulation [14, 15] as well as cell filamentation and death [16, 17]. Moreover, changes in light scattering has been successfully used for monitoring and sorting the best clones in an E. coli culture capable of producing interferon or growth hormone [18]. The capacity of conventional flow cytometers to measure intrinsic and/or extrinsic fluorescence in cells has been used as a property to assess cell viability, protein identity, and enzymatic activity [10, 11, 19]. Discussions regarding the specific properties of fluorescent probes have been published elsewhere [10–12]. Moreover, a complete list of fluorescent dyes can be found in the catalog of Molecular Probes [20]. Readers can find the most appropriate dye for specific FCM needs using these resources, including the target, the type of cell under study, and the excitation/emission wavelengths of the fluorescent compound. One step that may be problematic for FCM analysis during staining is that permeabilization and fixation may affect the viability of the cells; specific examples of how these problems can be overcome have been previously published [16]. Gene expression analysis in bacteria has been assessed by FCM employing fluorescent proteins including theAequorea green fluorescent protein (GFP) [21]. Special forms of GFP have been specifically tailored for its application in FCM [22, 23].

The small size of the bacteria could pose a limitation for FCM analysis specifically for the difficulty of distinguishing between small cells and cellular debris. To overcome this, the literature recommends using both forward scatter and fluorescence as dual trigger signals [10, 11]. Moreover, nucleic acid staining for distinguishing between cells and abiotic particles and the employment of polymer beads for standardization have been found useful in overcoming this limitation [16, 17, 24]. Finally, cell aggrupation, including the formation of bacterial chains or clusters of cells, may be problematic during FCM analysis since this technique measures events and cannot distinguish between a single cell and groups of cells passing through the interrogation point. Therefore, it is necessary to disaggregate and homogenize the sample before analysis; mild sonication has been applied to accomplish this purpose [25–27].

Fluorescence activated cell sorting (FACS) is an invaluable tool for separating the subpopulation(s) of interest that possess certain measurable characteristics. A cell sorter is basically a flow cytometer that has the option of separating cells. When the cell sorter detects a particle with the characteristics chosen by the operator, a charge is given to the droplet where that particle is, and as the droplets are passing through an electrostatic field formed between two charged plates, charged droplets are deflected appropriately, whereas uncharged droplets continue on their original course to the waste tank. With this methodology, around 40 000 cells can be separated per second [28]. As shown in Figure 2, what makes this technique relevant is the possibility of separating subpopulations for subsequent physiological and/or molecular analysis [29, 30]; moreover, cells can be recovered alive for growth and physiological analysis purposes [31]. However, if the cell population of interest exists in a very low abundance, it may be very difficult in principle to set the correct gate for sorting it. Secondly, it may take a long time to obtain a useful quantity of cells for downstream analysis; for example, at least 1 000 to 10 000 cells may be needed for obtaining detectable PCR products [30], and if the population of interest constitutes 0.1% of the total sample, it would take us approximately 4 minutes to obtain 10 000 cells at a sorting rate of 40 000 cells per second, and it would take us up to 41 minutes to obtain the sufficient number of cells if the abundance of such population is as low as 0.01%. Furthermore, to perform proteomics, up to $5 \times 10^6$to $10^9$ bacterial cells may be required, and it took 3 days and 3 weeks, respectively, to collect that number of cells of interest [29, 32, 33]. Both the time and gating problems have proven to be diminished by repeated sort cycles, enriching first the population of interest using a wide gate and after that using a narrower gate to purify it [30]. Otherwise, if we are interested in certain subpopulation that possess a transient physiological state for further analysis, the lapse of time required for separate those cells could be too long that the transient phenotype could be missed before reaching the number of cells needed.

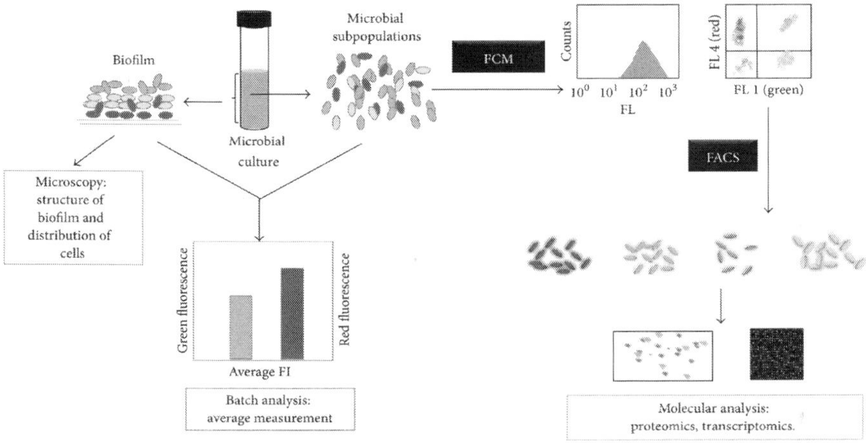


**Figure 2:** Left. Structure and cell distribution in biofilms can be studied by microscopy.Bottom left. Analysis of a bacterial culture sample allows obtaining average measurements of a physiological response. Right. FCM allows analyzing subpopulations of cultured bacterial cells with different physiological states; FACS permits separating those subpopulations that can be independently characterized by high throughput molecular techniques.

Although FCM emerged from 1960's as a tool for analyzing blood cells, advances in flow, optics, electronics, and probing techniques have done this approach suitable for study of a great number of purposes in different biological systems [12]. Quixabeira et al. [34] made an analysis of published literature regarding FCM applied to genetic studies, from 1991 to 2007, and they reported that more than a half of these studies are human related, the majority are related to diseases, and just a small proportion has been devoted to studying bacterial responses (viruses/bacteria, 8.4%). The main factor that has limited the use of FCM in prokaryote field is the small size of bacteria which makes them difficult for detection [4, 34, 35]. However, current improvements have allowed applying this technique to characterize the physiological events underlying the bacterial responses as described below.

## Flow Cytometry Analysis of Bacterial Response to Antibiotic Agents

Because an increasing number of resistant strains are emerging [36, 37], huge efforts are dedicated to new antibacterial drug research [38, 39] using automated methods to assess the number of candidates required, besides the necessity to find in terms of hours rather than days the appropriate drug for treating bacterial infections in clinical practice [40, 41].

When studying response to antibiotic agents, susceptibility is traditionally assessed in bulk, by the growth capability of the bacteria. It is possible to study viability as a parameter of antibiotic susceptibility at single-cell level, measuring criteria such as impermeability of membrane to dyes, maintenance of membrane potential, and the presence of metabolic activity measured by the production of a fluorescent metabolite from a nonfluorescent precursor [12].

Walberg et al. [24] employed FCM to assess the susceptibility of growing E. coli cells to mecillinam and ampicillin by using DNA staining with a combination of mithramycin and ethidium bromide (first reported by Steen et al., 1994 [42]). Prior to staining, the cells had to be permeabilized with ice-cold ethanol treatment. Fluorescence was recorded to measure DNA content, forward scatter recorded as cell size indicative, and side scatter as proportional to dry weight or protein content. These three parameters were augmenting in cells exposed to antibiotic since 30 minutes (and doubled at 60 min) incubated with minimal inhibitory concentration (MIC) of mecillinam and ampicillin. Thus showing that DNA and protein synthesis continued but cells were unable to complete division, since cell number remains constant. Such results were in agreement with the known mode of action of penicillin, which interferes with cell wall synthesis. Walberg's results indicate that, under the conditions tested, drug responses can be detected by light scatter alone, without staining the cells, but, as Walberg points out, in clinical samples, fluorescence measuring is important in order to distinguish between cells and other particles like debris. In an ongoing study [16], the same group demonstrated the applicability of flow cytometry measurement of DNA content and light scattering to assesses the susceptibility of antibiotic drugs, with different modes of action, including, ceftazidime a beta lactam that works as a cell wall antagonizer, ciprofloxacin, a quinolone that targets DNA gyrase, and gentamicin, an aminoglycoside that irreversibly binds to ribosomes. Their results, supported by microscopic observations, were consistent with the action mode of the tested drugs. They reported for these drugs that the cell size/dry weight ratio and DNA content augment as a function of drug concentration and exposure time. The same was true in fluorescence versus forward scatter plot with the existence of populations of filamenting cells, disintegrating filaments, and debris after treatment with the drugs. In a subsequent study, Walberg et al. [17]

reported the applicability of the previously exploited technique in studying a heterogeneous drug response, to a mixed culture of two clinically important bacteria species: E. coli and Klebsiella pneumoniae. These microorganisms are associated with polymicrobial urinary tract infection and both have different susceptibility to the drug tested, ampicillin [17]. This study was found to be valuable in detecting polymicrobial infections; by measuring cell number, light scattering, and DNA content-associated fluorescence, the findings show that it was possible to detect susceptible and resistant cells within an hour of incubation with ampicillin in the same sample.

In another study, Dessus-Babus et al. developed a flow cytometric method to calculate the minimal inhibitory concentration (MIC) to distinct antibiotics in clinical isolates of Chlamydia trachomatis and a reference strain [43]. They infected McCoy cells and later treated them with increasing doses of doxycycline, ofloxacin, and erythromycin; after incubation with the antibiotic, they immunostained McCoy cells with a fluorescein isothiocyanate- (FITC-) conjugated antibody to detect Chlamydia inclusions. Dessus-Babus and colleagues were able to quantify inclusion forming colonies (IFC)/mL by microscopy and by FCM; the latter method was not as sensitive as microscopy, but it did have the advantages of being specific and reproducible. Although infected McCoy cells were quantified instead of bacterial cells, this study demonstrated that flow cytometry is useful because the interpretation of results did not depend entirely on a skilled and experienced observer. Moreover, since FCM is an automated method, it is time-saving and results can be statistically improved.

Another application of flow cytometry to determine MIC of antibiotic drugs was reported by Assunção et al. in 2007. These authors succeeded in determining the MIC of nine different antibiotics in Mycoplasma hyopneumoniae, quantifying the total number of cells by means of staining DNA with SYBR Green. The results obtained with this approach were comparable with those obtained by the classical broth dilution method, which is dependent on the observation of a color change or increasing turbidity in the growth medium. However, the FCM approach was faster (12 h) than the broth dilution method (48 h), and furthermore, it was concluded that at 48 h FCM was more sensitive for tylosin and at 72 h for oxytetracycline and streptomycin MIC determination [44].

Recently, Soejima et al. reported for the first time an FCM based methodology to distinguish between live-injured and dead Listeria monocytogenes cells after treatment with antibiotics [45]. Furthermore, these authors were able to identify both states of bacterial cells in clinical blood samples, revealing the potential value of this approach to the opportune evaluation of bacteremia and the assessment of drug treatment in real time. This methodology employed photoactivated ethidium monoazide (EMA) that cleaves DNA of injured, not living cells, in combination with SYTO9 that enters live and dead cells and PI that penetrates dead cells that had lost membrane integrity [45].

Besides susceptibility to antibiotic assessment, relevant information can be obtained by FCM analysis in order to establish antibacterial drug action mechanisms [46–51]. Such information cannot be obtained by traditional culture-based techniques [52]. The use of fluorescent probes to detect specific cell changes [12], including, permeabilization and changes, in membrane potential [53–55], DNA content [16, 17, 24], and metabolic activity [53, 55], are useful parameters to assess viability and thus antibiotic susceptibility. Thus, Suller and Lloyd [52] performed a study to evaluate the effects of ceftazidime on Pseudomonas aeruginosa, ampicillin on E. coli, and vancomycin on Staphylococcus aureus, using the fluorescent probes bis(1,3-dibutylbarbituric acid), trimethine oxonol (DiBAC$_4$(3)), and SYTOX Green to measure membrane potential, the redox dye cyano-2,3-ditolyl tetrazolium chloride (CTC) to measure actively respiring bacteria, and the Baclite viability kit (Molecular Probes, Life Technologies, Grand Island, NY) to test viability. Results showed that the use of these fluorophores is effective to assess the antibacterial activities of the drugs tested although different responses between dyes were apparent. Thus, CTC was more efficient in detecting different subpopulations than DiBAC$_4$(3) and SYTOX Green when fluorescence values were plotted as a function of forward scatter; moreover, when the dyes (SYTO9 and propidium iodide) of the Baclite viability kit were used, no populations of differently responding cells were detected. Additionally, the use of these dyes has the advantage that no pretreatments of cells were required. Other studies have succeeded in employing fluorescent probes to measure the efficiency of new antibacterial drugs [12, 56, 57]. For instance, Ghosh et al. [57] demonstrated by FCM the antibacterial activity of two bioactive compounds isolated from seeds of Alpinia nigra against seven pathogenic bacteria. Ghosh and colleagues found by measuring of the extent of propidium iodide DNA staining of treated cells that both compounds caused significant increase in fluorescence intensity compared to the controls, which indicates membrane damage. The interpretation of this finding was supported by field emission scanning electronic microscopy;

they observed membrane disintegration and significant damage to the cell wall [57].

Since multiparametric flow cytometric analysis gives precise, reproducible, and accurate information regarding cell function at a single-cell level, the validity of growth based methods to assess cell viability has been questioned [58]. However, it has also been questioned whether cytometric measurements correlate with cell viability in the culture [12]; thus, it is reasonable to emphasize that there is no a universal formula that can be applied to every bacteria and every drug, but it is clear that FCM is a very sensitive, accurate, and time-saving method for assessing bacterial responses to antibiotics [16, 50]. Inconsistencies between cytometric analysis and growth based approaches can be explained by the existence of metabolically active but noncultivable cells (VNBC) in a cell culture [58, 59]. In fact, FCM has been successfully applied to detect VBNC cells from a variety of sanitary and clinically important sources, when the isolation and growth of viable cells are not possible [60, 61]. Cell sorting (by FACS) and further analysis of cells allows us to do further tests in order to corroborate the cytometric results with traditional methods, such as growth in culture and high-resolution single-cell microscopic analysis [58, 62].

As noted above, GFP reporter fusions are very useful for studying gene expression, and Sánchez-Romero and Casadesús [63] quantified GFP fluorescence by flow cytometry, in a liquid isogenic culture of Salmonella enterica expressing a ompC:gfp reporter gene fusion. In this study, heterogeneity was observed in the expression levels of the outer membrane porin coding gene; moreover, after sorting populations with high and low expression levels and assessing the susceptibility of such cell populations to kanamycin, it was found that low levels of ompC expression correlate with high kanamycin resistance. They proposed that the noisy expression ofompC is a mechanism contributing to the adaptive resistance to lethal concentrations of kanamycin [63]. In another study, Cui et al. used a GFP fusion as reporter of graF expression; the researchers sorted S. aureus cells with different levels of fluorescence by FACS, further susceptibility analysis to glycopeptide antibiotics, and morphology studies of graF overexpressing cells were performed. They found that upregulated activity of graFpromoter is consistent with reduced susceptibility to glycopeptides due to an increased cell wall thickness [64]. Another approach to detect resistant bacteria to antibiotics is using fluorescent drug analogs called reporter enzyme fluorescence [65, 66] or using drug fluorescent analogous with FRET pairs (fluorescence resonance energy transfer), that upon enzyme cleavage, the quenching molecule is eliminated and the fluorophore then emits light [67, 68]. An example of this approach was reported by Shao et al. who designed optical probes for detecting β-lactamase activity and covalent fluorescent labeling of antibiotic-resistant bacteria [69]. Therefore, an increasing number of future applications in research and clinical fields using the above-mentioned techniques may be anticipated, which can be substantially potentiated by employing personalized probes or dyes for monitoring particular targets and cellular functions.

## Measurement of Bacterial Responses to Other Chemical and Physical Stresses

Together with the search for chemical agents that are useful to kill bacteria, other strategies have also considered the idea of causing cell stress and stopping bacterial growth. One of such approaches employs photodynamic inactivation (PDI), which consists of the use of a nontoxic photosensitizer that is activated by harmless visible light. This then produces reactive oxygen species that causes fatal damage to target bacteria [70]. This approach has been tested in vitro [71] and in vivo [72]. Since PDI causes nonspecific injuries to the pathogens, it is very unlikely that bacteria can acquire resistance. One work related to ROS production by photoactivation of hypocrellin B (a component of a traditional Chinese herb H. bambuase) was reported by Jiang et al. [73]. In this report, clinically isolated S. aureus cells were incubated with hypocrellin B and after photodynamic treatment they used DCFH-DA (dichlorodihydrofluorescein diacetate) to measure by FCM the production of ROS [74]. ROS levels increased substantially in cells treated with PDI. The treatment caused reduction of viability as demonstrated by CFU counts and cellular damage, including membrane damage and cytoplasm leakage that was corroborated by confocal laser scanning microscopy [73].

FCM has also been applied in assessing the effectiveness of the treatments that food products for human consumption receive to eliminate and prevent pathogen dissemination and spoiling bacteria, including Listeria monocytogenes. This bacterium possesses a remarkable adaptability to stress conditions during food processing such as acidic pH, high salt concentration, and extreme temperatures [75]. It has been shown that, in order to cope with these stresses, this microorganism activates the expression of genes from the general stress sigma B ($\sigma^B$) regulon. To clarify the role of $\sigma^B$ during adaptation to low temperature, Utratna et al. analyzed the expression of a reporter gene fusion between the promoter of the $\sigma^B$ -dependent Imo2230 gene and GFP at normal and low temperatures. The FCM analysis revealed a heterogeneous activation of the $\sigma^B$ -dependent GFP fusion expression occurring since the early exponential phase, although the maximal expression occurred when cells entered into the stationary phase, of growth; however, such a result was found to occur at both temperatures tested. Therefore, they concluded that $\sigma^B$ does not play a pivotal role in adaptation to cold temperatures [76].

Another bacterium receiving a lot of attention in regard to stress response characterization, being a causal agent of food borne diseases, is Bacillus cereus [77]. Flow cytometry among other techniques has been useful in studying B. cereus responses to several chemical and physical stresses, such as low pH, which are encountered by the bacterium in several situations including the treatment that alimentary products are subjected to [78], in the host's gastrointestinal system [77, 79], and to other sporistatic/sporicidal physicochemical treatments [80, 81].

FCM has also been applied to investigate the way in which bacteria respond and adapt to stressful conditions that they face during industrial and remediation bioprocesses [82]. Streptococcus macedonicus, Bacillus licheniformis, and Lactobacillus rhamnosus are examples of bacteria employed in industrial fermentations. Papadimitriou et al. [55] reported the use of FCM to assess in situ the physiological status of S. macedonicus in response to acid stress. This microorganism that is a member of the lactic acid bacteria group is widely used in food industry. In this report, Papadimitriou and colleagues employed FCM to analyze characteristics such as membrane potential with $DiBAC_4(3)$, membrane integrity with propidium iodide, and enzymatic activity as well as membrane integrity with cFDA, carboxyfluorescein diacetate. Finally, they determined cultivability after cell sorting. From these analyses, the coexistence of three distinct subpopulations was observed: intact/culturable, permeabilized/dead, and potentially injured with decreased culturability [55].

Sunny-Roberts and Knorr monitored changes in membrane integrity using propidium iodide and esterase activity with carboxyfluorescein diacetate of L. rhamnosus in response to osmotic stress, which is a condition that this bacterium has to challenge in natural habitats as well as in food formulations and processes where a probiotic bacterium is used. By FCM and conventional culture techniques, it was found that this microorganism is able to tolerate even extreme sucrose concentrations [83], and such a characteristic is exploitable for its use in food processing and formulations.

In another study, FCM was employed to assess physiological responses of the industrially important bacteriumB. licheniformis, used in bioremediation processes. Using the same technique, researchers have monitored population dynamics of continuous cultivations, responses to starvation conditions, and responses to glucose and lactose pulses [84, 85].

The heterologous expression of proteins in E. coli is a routine method for obtaining large amounts of recombinant proteins. As this process may represent a stressful condition to E. coli cells, Borth et al. analyzed by FCM the physiological changes in response to the metabolic stress that represents the production of foreign proteins at high yield. They measured total DNA and RNA content (indicators of cell division and protein synthesis activity, resp.), total protein content (cell size), and intracellular SOD content (foreign protein). They found that after induction of foreign protein synthesis, cells increased their biomass but did not divide. Thus, they conclude that it is important to fine-tune the expression system in order to prolong lifetime and therefore production yield [86]. Thus, FCM could help monitoring physiological changes in response to fine-tuning. Trip et al. developed an elegant, FCM-FACS based system to detect and separate B. subtilis cells that secrete heterologous proteins in large amounts. This system is proposed to be used in screening for biotechnologically important secreted proteins from genomic libraries of uncharacterized bacteria. Such a system consists of aphtrA-gfp reporter system, whose expression is activated by the CssRS secretion stress response; thus, when the reporter B. subtilis strain over produces and secretes proteins, the CssRS system is activated and promotes the expression of gfp under control of htrA promoter, thus producing a green florescent B. subtilis cell that can be separated by FACS for further propagation and analysis [87].

FCM has been also employed to characterize the physiological response of bacteria to organic compounds, such as phenol and toluene. A study that aimed to investigate these points was carried out by Wiacek et al. [33]. It was observed that a culture of Cupriavidus necator exhibited a heterogeneous response to harmful phenol concentration treatment determined by measuring chromosomal DNA and PHB content (with DAPI and Nile red, resp.), and after sorting different responding subpopulations they performed functional proteomics analysis [33]. The same group also investigated later the mechanism by which some bacteria uptake toluene. By means of a fluorescent labeled toluene analogue dye (NBDT), they measured toluene uptake using FCM, and after cell sorting, proteomics analysis revealed the presence of toluene transporting porins in P. putida mt-2when grown on toluene but not when grown in glucose [88].

The latest works exploited the advantage that cell sorting offers, which includes further analysis of subpopulations, either for obtaining more accurate results than those obtained by analyzing batch cultures or for comparing and characterizing the physiological state of different subpopulations.

## Host-Pathogen Interaction

FCM is an automated technique suitable for studying bacterial interactions with host cells, useful for evaluating exclusively the interaction process or even for characterizing such interaction.

Analysis of expression of GFP reporter fusions during analysis of host-pathogen interactions is one of the most common methods employed; following this approach neither the physiology nor the pathogenesis course is altered [89–93].

Stapleton et al. established and standardized the use of FCM to analyze the relative adherence of uropathogenicE. coli to clinical samples of exfoliated vaginal epithelial cells (VEC). The researchers employed three GFP-expressing E. coli strains isolated from women with urinary tract infections, expressing different types of fimbria and with different papG classes. They counted PI stained VEC's, measured adhered bacterial associated GFP fluorescence, and determined the percentage of positive VEC's and the number of adhered bacterial cells. Their results showed different adherence patterns among E. coli strains expressing different adhesins [94].

Hara-Kaonga and Pistole's study is another example of the application of FCM to detect bacteria-host cell interactions. They were interested in clarifying whether ompC and ompD from Salmonella enterica were involved in the recognition by macrophages and epithelial cells. Specifically, the researchers analyzed whether these cells differentially recognize single and double ompC ompD null mutants. To this end, bacterial cells were stained with FITC and were later incubated with macrophages or epithelial cells. Then, the fluorescent human cells were detected by FCM and the number of adhered bacterial cells was determined. Their results showed that OmpD, but not OmpC porin, is involved in the recognition of S. enterica serovar typhimurium by human macrophages and epithelial cells

[95]. The same authors later reported a method for assessing the adherence ofS. enterica to host cells by measuring FITC associated fluorescence of bacteria and lipophilic dye PKH-26 associated fluorescence of eukaryotic cells, therefore eliminating the inaccuracies due to bacterial cell aggregates that mimic forward scattering of eukaryotic cells [96].

Another application of FCM in the study of pathogen-host interaction is analyzing genes that are specifically expressed when this interaction takes place. An example was reported by Bent et al. who investigated the expression of yspP and orf6 genes, codifying, respectively, an effector protein secreted by and a structural component of the T3SS (type 3 secretion system) apparatus. The yspP and orf6 expressions were determined by GFP reporter fusions to those gene's promoters. The reported Yersinia enterocolitica cells were used to infect mice. Infected murine cells were subjected to flow cytometric analysis in order to determine GFP fluorescence associated to yspP and orf6 genes expression; this approach combined with qRT-PCR demonstrated that Ysa T3SS is expressed in infected mice [97].

With the purpose of separating cells of interest for further analysis using FACS technology, reports have been published that analyze gene expression during host-pathogen interaction. One common approach was to make a fusion library of chromosome fragments and to clone the fragments in a vector carrying a promoterless gfpgene to capture promoters. The approach is the following: bacterial cells transformed with plasmids that express GFP when interacting with host cells are separated by FACS and are later recovered, and their promoter is then at least partially sequenced to identify genes expressed during host-pathogen interaction. With this experimental approach, Barker et al. investigated the genes of Mycobacterium marinum that are differentially expressed during macrophage phagocytosis. They sorted fluorescent phagosomes, and after 2 to 3 days after infection, they separated vesicles containing single cells; following this approach, they identified 12 clones containing GFP fusions differentially expressed including membrane proteins and biosynthetic enzymes [98]. Following a similar approach, Wilson et al. analyzed the in vivo transcriptional response when Listeria monocytogenes infected a murine macrophage model [99].

Thus, flow cytometry offers a rapid, accurate, and reproducible method to quantify the number of wild type or mutant bacterial cells adhered or phagocyted by host cells, with technical complications that each bacterium and host cell type represents. On the other hand, the expression of selected genes during host-pathogen interaction can be assessed, or even genes participating in infection processes can be identified using FCM coupled to cell sorting for further analysis. Although fluorescence microscopy could be an alternative for these purposes, FCM offers the advantage of analyzing and counting a high number of cells, thereby improving the statistics. However, it must be pointed that FCM does not provide information regarding the distribution or location of bacteria within the host cell, and hence a microscopic analysis could be an excellent complement to this technique.

The pathogenic bacteria Shigella flexneri are capable of causing dysentery by invading the epithelial cells of the colon. To identify the factors that allow this bacterium to infect the epithelial layer, an FACS based approach was employed to sort mutant clones incapable of spreading within and between epithelial cells. At least three different classes of mutants were identified with this approach, namely, those that presented an altered lipopolysaccharide structure, clones that were affected in intracellular motility, or those that exhibited defect in its capacity to invade the cell hosts [100]. With the purpose of understanding the molecular basis of bacterial virulence, genetic approaches combined with FCM have been applied to identify and isolate nonvirulent bacteria. In a recent study, a liquid culture of Vibrio cholerae was subjected to random transposon mutagenesis with the purpose of generating mutants deficient in virulence activation. The entire population was subjected to FACS to successfully separate the cells that were deficient for virulence activation [101]. In another report, FACS was applied to separate recombinant Mycobacterium capable of expressing high levels of a foreign ovalbumin epitope; notably, the sorted clones were more efficient in inducing an immunogenic response. These studies demonstrated that recombinant Mycobacterium has the potential to be employed as vehicles to delivering pathogen antigen peptides and inducing systemic and mucosal immune responses [102].

## Analysis of Bacterial Physiology at a Single-Cell Level

FCM has also been applied to understand the physiological changes occurring in bacterial cells in response to environment changes. One of the most regulated processes in bacteria is cell division. In Gram-positive bacteria, the set of proteins required for cell division is known; these cell division proteins (CDPs) must act in a specific yet known order. Trip et al. were interested in investigating if transcription of those CDPs is the underlying process that regulates speed of cell division. To investigate this aspect, transcriptional gene fusions between CPD promoters and the GFP encoding gene were recombined into a neutral locus of the chromosome of B. subtilis. The GFP-fusion containing strains were employed to measure by FCM and qRT-PCR if CDPs expression varies or not under different culture conditions that modify growth rate. Constant expression levels of the fusions were found when tested independently of the growth rate and cell cycle; therefore, it was concluded that cell division in B. subtilis is regulated mainly at a posttranslational level and is influenced by other factors that do fluctuate, such as metabolic state and substrate availability, but not by transcription [103].

DNA staining and subsequent FCM quantitative analysis has been applied to gain insights into mechanisms that underlie cell cycle regulation and chromosome biology in bacterial cells. Such knowledge impacts on the understanding of the life cycle of the bacterial cell and can be used to determine cell states quickly and thus predicts the metabolic and survival behavior of microorganisms [4, 104]. DNA quantification patterns are indicative of the number or chromosomes, and by FCM analysis information can be obtained about the number of individual cells that possess an n number of chromosomes in an asynchronous growing population. These patterns are characteristic of distinct bacterial species and change under different environment conditions (reviewed by [105]). When studying protein production, function, and localization in batch cultures, it is important to synchronize the cells. FCM analysis of DNA stained with chromomycin has been used to ensure the quality of synchronous population of Caulobacter crescentus for studying morphogenesis and cell cycle regulatory proteins that function along the cell cycle [106]. DNA staining histograms obtained by FCM, after treating the cells with drugs that inhibit replication initiation and cell division, are an approach employed to find the number of chromosomes per cell. Thus, variations in these distribution patterns may be analyzed as a function of different conditions and genetic backgrounds to gain insights into cell cycle regulation [107, 108].

There is also a wide interest in studying bacterial differentiation due to the impact of resistant spores and antibiotics resistance related to competence development over clinical, sanitary, and biotechnological issues. B. subtilis is a suitable model for studying the molecular mechanisms underlying natural competence development and sporulation [109]. Both developmental pathways are interconnected and appear to be mutually exclusive. Chung et al. (1994) published one of the pioneering works that applied FCM to investigate the heterogeneous development of spore differentiation pathways in B. subtilis. In this work, using lacZ reporter fusions with early spore development genes spoVG, spoIIG, and spoIID, they measured β-galactosidase activity at the single-cell level, revealed by the hydrolysis of the fluorogenic compound C8-FDG (5-octanoylaminofluorescein-di-β-D-galactopyranoside) that releases C8-fluorescein which emits green light when excited at 488 nm. They found with this approach that, within a culture of sporulating B. subtilis cells, there are two distinct subpopulations and only one of them initiates and concludes with the spore developmental program [110]. FCM and fluorescent reporter fusions to promoters have been employed to gain insights into the pathways that control sporulation and competence development and in the molecular mechanisms by which these processes occur in heterogeneous populations within an isogenic culture; this is important because only a small fraction of cells may experience those developmental processes. Smits et al., employing a (ComK activated) $P_{comG}$-gfp to analyze competence development at single-cell level by FCM, showed the importance of the transcriptional autostimulation of comK in the development of competence and also demonstrated that this state occurs in a population of cells that reach a threshold level of ComK [111]. Later, following a single-cell analysis approach, that analyzed the expression of a $P_{spoIIA}$-gfp fusion activated by Spo0A, allowed them to conclude that the autostimulatory activation of Spo0A is responsible for the bistable expression pattern in sporulating cultures [112]. Using the same reporter fusion, this research group performed single-cell analyses on the expression patterns of both competence development and sporulation of a B. subtilis culture of a 1 : 1 mix spore induction: competence induction in chemically defined liquid media and also in biofilms. They found that, in the 1 : 1 mix, both processes are sequentially initiated; first competence develops and sporulation activates later. In another experimental approach, it was found that a small fraction of spores were also capable of acquiring competence. They also found that sporulation is more effectively initiated in biofilms than in planktonic cells. Moreover, this group reported that, under conditions that do not usually trigger sporulation or competence, there are few cells that form spores or that become competent respectively; that is, both differentiation pathways are noisy. Of note, these analyses were only possible by employing FCM noise measurement, since at least 100 000 cells must be analyzed under rigorous gating to detect positive cells [109]. Later, this group identified RapH as a novel factor involved in the temporal separation of competence and sporulation. They observed, by using approaches such as single-cell analysis, that the overproduction of RapH provokes a drastic reduction of competence and sporulation gene expression and that rapH genetic disruption causes a significant increase in the frequency of cells simultaneously expressing both sporulation and competence fusion reporters; therefore, in a rapH mutant sporulation initiation and competence development are no strictly separated [113]. The mechanisms governing the production of exoproteases in nonsporulating cells of B. subtilis, within heterogeneous populations, were analyzed by FCM using strains harboring GFP reporter fusions. It was found that only a fraction of the vegetative cells turned on the expression of the bacillopeptidase (bpr) and subtilisin (aprE) encoding genes under control of the regulator DegU [114].

In a different bacterium, Stecchini et al. [8] investigated the effects of changes in humidity and viscosity in growth cultures of B. cereus, a human pathogen [77] over cell motility, spore dimension, and thermal resistance. Such parameters were measured by forward and side scattering (assess dimension) and propidium iodide (thermal damage) [8].

Cronin and Wilkinson, using differential staining, microscopy, and FCM analyses, reported the establishment of a new methodology to identify, quantify, and assess changes in permeability and metabolism of germinatingB. cereus endospores. This methodology consisted in FCM single-cell level measurement of CFDA/Hoechst 33342 to estimate overall germination rate; and by measuring side-scatter and SYTO9 staining, they quantified ungerminated, germinating, and outgrowing endospores [115].

The list of discoveries enriched and supported by cytometric analysis related to physiological bacterial functions is still growing. Another interesting issue is quorum sensing (QS), a way by which bacteria communicate by secreting and responding to signal molecules or autoinducers. Such processes are relevant because in pathogenic bacteria certain virulence factors encoding genes are regulated by quorum sensing [116]. Using FCM analysis of reporter strains expressing promoter-gfp fusions, Anetzberger et al. monitored the induction/repression of autoinducer-regulated genes in the shrimp pathogen Vibrio harveyi. It was found that, besides luminescence, exoprotease gene expression is also regulated by cell density and evidenced by single-cell analysis of simultaneously assessed luminescence and exoprotease expression which they observed functional heterogeneity within the population [117].

In nature, most bacteria exist as aggregates surrounded by an extracellular matrix that is in a biofilm [118]. Biofilms are conformed by specialized subpopulations, with different physiological characteristics compared to those exhibited as planktonic cells [25, 26, 119]. In several contexts, including clinical, industrial, and environmental ones, there is a great interest in understanding how a biofilm is formed and how the cells composing it behave [119].

Given the nature of the biofilm, microscopy has been used as the choice method to analyze the structure, localization, or distribution of the biofilm components; however, FCM has also been successfully utilized for analyzing the physiological conditions of the cells in a biofilm [25]. A requirement for FCM analysis is that cells must be free, in a suspension; to accomplish this, the biofilm has to be disrupted, for instance by passing it by trough a pipette or a needle or by mild sonication [25, 120].

The importance of heterogeneous subpopulations in biofilm formation, using B. subtilis as model, has been investigated by FCM. Using this technique, it was discovered that the specialized matrix which is essential for biofilm formation is composed by TasA amyloid-like fibers, exopolysaccharide, and hydrophobin BslA and is produced by a specific subpopulation [27, 121–125]. Garcia-Betancur et al. reported the use of fluorescence microscopy combined with FCM to visualize and quantify the subpopulations of matrix producers and surfactin secretors (signaling molecule that triggers differentiation of matrix producers) within biofilms of B. subtilis, with the use of fluorescent reporter fusions with promoters of genes required for matrix ($P_{tapA}$- CFP) and surfactin ($P_{srfAA}$-YFP) production. They found that the reporter fusions are expressed only in certain subpopulations, under biofilm formation induction, and in the case of the double-labeled strain showed a single population of fluorescent cells expressing both CFP and YFP, which indicates that in the same population both differentiation pathways are coordinately activated [25].

Using the same microorganism, Marlow et al. reported another type of specialized cells involved in biofilm formation. These authors applied FCM and microscopy to detect exoprotease-producing cells within the biofilm. To this end, fluorescent proteins fusions to $P_{bpr}$ as reporter of expression of exoprotease-encoding gene bpr and other fluorescent reporter fusion $P_{tapA}$-mKate2 to monitor matrix producing cells were employed. It was found that the number of exoprotease producing cells increased as the biofilm matures and that exoprotease production is dependent on the levels of the response regulator phosphorylated DegU (that controls swarming motility, biofilm formation, and exoprotease production); moreover, it was found that exoprotease producing cells arise from both matrix-producing and nonproducing cells. Using microscopic analysis of a cross-section of a mature biofilm, they found that the subpopulation that produced exoprotease is enriched in the air interface more than that in the agar interface [126].

Although the above-mentioned works were performed in vitro, there are also in vivo studies; for example, Beauregard et al. demonstrated that B. subtilis colonizes Arabidopsis thaliana roots forming biofilms. This colonization has beneficial consequences for plant development. Although the usefulness of this bacterium as a biofertilizer is well known, Beauregard's results established the importance of biofilm formation in plant colonizing. They determined by FCM that plant polysaccharides, like arabinogalactan, pectin, and xylan, play signaling roles during biofilm formation and as a source of sugars for the synthesis of extracellular matrix. This was made by measuring the levels of fluorescence of biofilm forming cells expressing the $P_{tapA}$-yfp fusion [127]. These results were consistent with in vitro pellicles formation assays and in vivo biofilm formation as was observed by fluorescence microscopy.

B. licheniformis is a useful bacterium for treating waste residues in water generated from alimentary industrial processes and is therefore important to optimize culture conditions for improved cell aggregation and biomass separation. da Silva et al., using FCM combined with biomass quantitation and confocal microscopy, established such optimal conditions, by assessing changes in the limiting nutrient, dilution rate, and agitation intensity looking for those that gave the cell aggregates where the majority of cells were metabolically active [128].

The employment of FCM in studying biofilms is very useful but not sufficient. Although this technique contributes to the statistics and accuracy of determinations, at single-cell level, it does not provide information regarding the structure in the biofilm and distribution of the distinct subpopulations.

## CONCLUDING REMARKS

FCM is a tool that is currently applied to bacterial analysis from detecting and counting bacteria, to determining changes in cellular functions and metabolic activity, and even in identifying genes that are expressed specifically under certain conditions. Although these analyses can be applied to a sample of cultured bacteria following the classic dilution and plating based methodology, FCM offers real measurements for each cell assessed, and in less time. With an appropriate combination of dyes, the damage caused by antibiotics and other threatening agents may be deduced and the proportion of affected cells determined; moreover, using fluorescent postenzymatic cleavage substrate analogs, metabolic activity can also be determined. FCM is an automated technique: it is time-saving, accurate, sensitive, and as thousands of cells can be processed per second, statistics are improved; this point is relevant and it is what makes FCM indispensable for certain purposes, such as identifying and quantifying cells present in very low abundance in a population. However, perhaps the most remarkable potential of FCM when coupled to cell sorting is that offers a unique opportunity to separate specific subpopulations of bacteria that exhibit differential physiological states, from entire cultures. Applying proteomics and/or transcriptomic analysis to characterize such subpopulations makes this approach a powerful tool to understand the cellular and molecular mechanisms that govern cell differentiation in bacteria. Finally, it must be pointed out that FCM and FACS require expensive equipment and skilled personal to operate it and interpret the results. With the appearance of low cost equipment in the market, it is expected that, in the near future, more studies can be carried out to study physiological responses in bacteria using these powerful approaches.

## ACKNOWLEDGMENTS

This work was supported by the University of Guanajuato (Grant no. DAIP-324-2013) and by the Consejo Nacional de Ciencia y Tecnología (CONACYT) (Grants nos. 205744 and 221231) of México. Verónica Ambriz-Aviña was supported by scholarships from CONACYT and the University of Guanajuato. The authors wish to acknowledge the support of DAIP-UGTO with professional editing assistance.

## REFERENCES

1. P. B. Rainey, H. J. E. Beaumont, G. C. Ferguson et al., "The evolutionary emergence of stochastic phenotype switching in bacteria," Microbial Cell Factories, vol. 10, article S14, no. 1, 2011.
2. D. Fraser and M. Kærn, "A chance at survival: gene expression noise and phenotypic diversification strategies," Molecular Microbiology, vol. 71, no. 6, pp. 1333–1340, 2009.
3. I. G. de Jong, J. Veening, and O. P. Kuipers, "Single cell analysis of gene expression patterns during carbon starvation in Bacillus subtilis reveals large phenotypic variation," Environmental Microbiology, vol. 14, no. 12, pp. 3110–3121, 2012.
4. S. Müller and H. Davey, "Recent advances in the analysis of individual microbial cells," Cytometry A, vol. 75, no. 2, pp. 83–85, 2009.
5. M. E. Lidstrom and M. C. Konopka, "The role of physiological heterogeneity in microbial population behavior," Nature Chemical Biology, vol. 6, no. 10, pp. 705–712, 2010.
6. D. Garmyn, L. Gal, R. Briandet et al., "Evidence of autoinduction heterogeneity via expression of the Agrsystem of Listeria monocytogenes at the single-cell level," Applied and Environmental Microbiology, vol. 77, no. 17, pp. 6286–6289, 2011.
7. B. Snijder and L. Pelkmans, "Origins of regulated cell-to-cell variability," Nature Reviews Molecular Cell Biology, vol. 12, no. 2, pp. 119–125, 2011.
8. M. L. Stecchini, M. Spaziani, M. D. Torre, and S. Pacor, "Bacillus cereus cell and spore properties as influenced by the micro-structure of the medium," Journal of Applied Microbiology, vol. 106, no. 6, pp. 1838–1848, 2009.
9. C.-C. Shu, A. Chatterjee, G. Dunny, W.-S. Hu, and D. Ramkrishna, "Bistability versus bimodal distributions in gene regulatory processes from population balance," PLoS Computational Biology, vol. 7, no. 8, Article ID e1002140, 2011.
10. H. M. Davey and D. B. Kell, "Flow cytometry and cell sorting of heterogeneous microbial populations: the importance of single-cell analyses," Microbiological Reviews, vol. 60, no. 4, pp. 641–696, 1996.
11. R. C. Leif, Practical Flow Cytometry, by M. Howard and M.D. Shapiro, Wiley-Liss, Inc, New York, NY, USA, 3rd edition, 1995.

12. H. M. Shapiro, "Microbial analysis at the single-cell level: tasks and techniques," Journal of Microbiological Methods, vol. 42, no. 1, pp. 3–16, 2000.
13. D. A. Veal, D. Deere, B. Ferrari, J. Piper, and P. V. Attfield, "Fluorescence staining and flow cytometry for monitoring microbial cells," Journal of Immunological Methods, vol. 243, no. 1-2, pp. 191–210, 2000.
14. P. Fouchet, S. Jan, J. Courtois, B. Courtois, G. Frelat, and J. N. Barbotin, "Quantitative single-cell detection of poly(β-hydroxybutyrate) accumulation in Rhizobium meliloti by flow cytometry," FEMS Microbiology Letters, vol. 126, no. 1, pp. 31–35, 1995.
15. F. Srienc, B. Arnold, and J. E. Bailey, "Characterization of intracellular accumulation of poly-β-hydroxybutyrate (PHB) in individual cells of Alcaligenes eutrophus H16 by flow cytometry," Biotechnology and Bioengineering, vol. 26, no. 8, pp. 982–987, 1984.
16. M. Walberg, P. Gaustad, and H. B. Steen, "Rapid assessment of ceftazidime, ciprofloxacin, and gentamicin susceptibility in exponentially-growing E. coli cells by means of flow cytometry," Cytometry, vol. 27, no. 2, pp. 169–178, 1997.
17. M. Walberg, P. Gaustad, and H. B. Steen, "Rapid discrimination of bacterial species with different ampicillin susceptibility levels by means of flow cytometry," Cytometry, vol. 29, no. 3, pp. 267–272, 1997.
18. K. D. Wittrup, M. B. Mann, D. M. Fenton, L. B. Tsai, and J. E. Bailey, "Single-cell light scatter as a probe of refractile body formation in recombinant Escherichia coli," Nature Biotechnology, vol. 6, no. 4, pp. 423–426, 1988.
19. H. B. Steen, "Light scattering measurement in an arc lamp-based flow cytometer," Cytometry, vol. 11, no. 2, pp. 223–230, 1990.
20. R. P. Haugland, Handbook of Fluorescent Probes and Research Products, vol. 1, Molecular Probes, Eugene, Ore, USA, 9th edition, 2002.
21. M. Chalfie, Y. Tu, G. Euskirchen, W. W. Ward, and D. C. Prasher, "Green fluorescent protein as a marker for gene expression," Science, vol. 263, no. 5148, pp. 802–805, 1994.
22. A. B. Cubitt, R. Heim, S. R. Adams, A. E. Boyd, L. A. Gross, and R. Y. Tsien, "Understanding, improving and using green fluorescent

proteins," Trends in Biochemical Sciences, vol. 20, no. 11, pp. 448–455, 1995.

23. M. Zimmer, "Green fluorescent protein (GFP): applications, structure, and related photophysical behavior," Chemical Reviews, vol. 102, no. 3, pp. 759–781, 2002.
24. M. Walberg, P. Gaustad, and H. B. Steen, "Rapid flow cytometric assessment of mecillinam and ampicillin bacterial susceptibility," Journal of Antimicrobial Chemotherapy, vol. 37, no. 6, pp. 1063–1075, 1996.
25. J. C. Garcia-Betancur, A. Yepes, J. Schneider, and D. Lopez, "Single-cell analysis of Bacillus subtilis biofilms using fluorescence microscopy and flow cytometry," Journal of Visualized Experiments, no. 60, Article ID e3796, 2012.
26. H. Vlamakis, C. Aguilar, R. Losick, and R. Kolter, "Control of cell fate by the formation of an architecturally complex bacterial community," Genes & Development, vol. 22, no. 7, pp. 945–953, 2008.
27. S. S. Branda, F. Chu, D. B. Kearns, R. Losick, and R. Kolter, "A major protein component of the Bacillus subtilis biofilm matrix," Molecular Microbiology, vol. 59, no. 4, pp. 1229–1238, 2006.
28. L. A. Herzenberg and R. G. Sweet, "Fluorescence-activated cell sorting," Scientific American, vol. 234, no. 3, pp. 108–117, 1976.
29. M. Jahn, J. Seifert, M. von Bergen, A. Schmid, B. Bühler, and S. Müller, "Subpopulation-proteomics in prokaryotic populations," Current Opinion in Biotechnology, vol. 24, no. 1, pp. 79–87, 2013.
30. G. Wallner, B. Fuchs, S. Spring, W. Beisker, and R. Amann, "Flow sorting of microorganisms for molecular analysis," Applied and Environmental Microbiology, vol. 63, no. 11, pp. 4223–4231, 1997.
31. T. U. Sørensen, G. J. Gram, S. D. Nielsen, and J. S. Hansen, "Safe sorting of GFP-transduced live cells for subsequent culture using a modified FACS vantage," Cytometry, vol. 37, no. 4, pp. 284–290, 1999.
32. N. Jehmlich, T. Hübschmann, M. Gesell Salazar et al., "Advanced tool for characterization of microbial cultures by combining cytomics and proteomics," Applied Microbiology and Biotechnology, vol. 88, no. 2, pp. 575–584, 2010.
33. C. Wiacek, S. Müller, and D. Benndorf, "A cytomic approach reveals population heterogeneity of Cupriavidus necator in response to harmful phenol concentrations," Proteomics, vol. 6, no. 22, pp. 5983–5994, 2006.

34. V. B. L. Quixabeira, J. C. Nabout, and F. M. Rodrigues, "Trends in genetic literature with the use of flow cytometry," Cytometry A, vol. 77, no. 3, pp. 207–210, 2010.
35. G. Nebe-von-Caron, "Standardization in microbial cytometry," Cytometry Part A, vol. 75, no. 2, pp. 86–89, 2009.
36. J. A. Heinemann, R. G. Ankenbauer, and C. F. Amábile-Cuevas, "Do antibiotics maintain antibiotic resistance?" Drug Discovery Today, vol. 5, no. 5, pp. 195–204, 2000.
37. B. Spellberg, R. Guidos, D. Gilbert et al., "The epidemic of antibiotic-resistant infections: a call to action for the medical community from the infectious diseases society of America," Clinical Infectious Diseases, vol. 46, no. 2, pp. 155–164, 2008.
38. P. S. Negi, G. K. Jayaprakasha, and B. S. Jena, "Antibacterial activity of the extracts from the fruit rinds of Garcinia cowa and Garcinia pedunculata against food borne pathogens and spoilage bacteria," LWT: Food Science and Technology, vol. 41, no. 10, pp. 1857–1861, 2008.
39. C. Gaspar-Marques, P. Rijo, M. F. Simões, M. A. Duarte, and B. Rodriguez, "Abietanes from Plectranthus grandidentatus and P. hereroensis against methicillin- and vancomycin-resistant bacteria," Phytomedicine, vol. 13, no. 4, pp. 267–271, 2006.
40. J. Barenfanger, C. Drake, and G. Kacich, "Clinical and financial benefits of rapid bacterial identification and antimicrobial susceptibility testing," Journal of Clinical Microbiology, vol. 37, no. 5, pp. 1415–1418, 1999.
41. H. M. Shapiro, "Multiparameter flow cytometry of bacteria: implications for diagnostics and therapeutics," Cytometry, vol. 43, no. 3, pp. 223–226, 2001.
42. H. B. Steen, M. W. Jernaes, K. Skarstad, and E. Boye, "Chapter 28 staining and measurement of DNA in bacteria," Methods in Cell Biology, vol. 42, pp. 477–487, 1994.
43. S. Dessus-Babus, F. Belloc, C. M. Bébéar et al., "Antibiotic susceptibility testing for Chlamydia trachomatis using flow cytometry," Cytometry, vol. 31, no. 1, pp. 37–44, 1998.
44. P. Assunção, N. T. Antunes, R. S. Rosales et al., "Application of flow cytometry for the determination of minimal inhibitory concentration of several antibacterial agents on Mycoplasma hyopneumoniae," Journal of Applied Microbiology, vol. 102, no. 4, pp. 1132–1137, 2007.

45. T. Soejima, J. Minami, and K. Iwatsuki, "The exclusive use of flow cytometry to evaluate the antibiotic-susceptibility," Biochimica et Biophysica Acta—General Subjects, vol. 1820, no. 12, pp. 1980–1986, 2012.
46. D. J. Mason, E. G. M. Power, H. Talsania, I. Phillips, and V. A. Gant, "Antibacterial action of ciprofloxacin," Antimicrobial Agents and Chemotherapy, vol. 39, no. 12, pp. 2752–2758, 1995.
47. B. L. Roth, M. Poot, S. T. Yue, and P. J. Millard, "Bacterial viability and antibiotic susceptibility testing with SYTOX green nucleic acid stain," Applied and Environmental Microbiology, vol. 63, no. 6, pp. 2421–2431, 1997.
48. M. T. E. Suller, J. M. Stark, and D. Lloyd, "A flow cytometric study of antibiotic-induced damage and evaluation as a rapid antibiotic susceptibility test for methicillin-resistant Staphylococcus aureus," Journal of Antimicrobial Chemotherapy, vol. 40, no. 1, pp. 77–83, 1997.
49. F. C. Mortimer, D. J. Mason, and V. A. Gant, "Flow cytometric monitoring of antibiotic-induced injury in Escherichia coli using cell-impermeant fluorescent probes," Antimicrobial Agents and Chemotherapy, vol. 44, no. 3, pp. 676–681, 2000.
50. D. J. Novo, N. G. Perlmutter, R. H. Hunt, and H. M. Shapiro, "Multiparameter flow cytometric analysis of antibiotic effects on membrane potential, membrane permeability, and bacterial counts of Staphylococcus aureus and Micrococcus luteus," Antimicrobial Agents and Chemotherapy, vol. 44, no. 4, pp. 827–834, 2000.
51. H. J. Wickens, R. J. Pinney, D. J. Mason, and V. A. Gant, "Flow cytometric investigation of filamentation, membrane patency, and membrane potential in Escherichia coli following ciprofloxacin exposure," Antimicrobial Agents and Chemotherapy, vol. 44, no. 3, pp. 682–687, 2000.
52. M. T. Suller and D. Lloyd, "Fluorescence monitoring of antibiotic-induced bacterial damage using flow cytometry," Cytometry, vol. 35, no. 3, pp. 235–241, 1999.
53. C. Laflamme, S. Lavigne, J. Ho, and C. Duchaine, "Assessment of bacterial endospore viability with fluorescent dyes," Journal of Applied Microbiology, vol. 96, no. 4, pp. 684–692, 2004.

54. J. Lehtinen, J. Nuutila, and E. Lilius, "Green fluorescent protein-propidium iodide (GFP-PI) based assay for flow cytometric measurement of bacterial viability," Cytometry A, vol. 60, no. 2, pp. 165–172, 2004.
55. K. Papadimitriou, H. Pratsinis, G. Nebe-von-Caron, D. Kletsas, and E. Tsakalidou, "Rapid assessment of the physiological status of Streptococcus macedonicus by flow cytometry and fluorescence probes," International Journal of Food Microbiology, vol. 111, no. 3, pp. 197–205, 2006.
56. I. Faria-Ramos, M. J. Espinar, R. Rocha et al., "A novel flow cytometric assay for rapid detection of extended-spectrum beta-lactamases," Clinical Microbiology and Infection, vol. 19, no. 1, pp. E8–E15, 2013.
57. S. Ghosh, K. Indukuri, S. Bondalapati, A. K. Saikia, and L. Rangan, "Unveiling the mode of action of antibacterial labdane diterpenes from Alpinia nigra (Gaertn.) B. L. Burtt seeds," European Journal of Medicinal Chemistry, vol. 66, pp. 101–105, 2013.
58. G. Nebe-Von-Caron, P. J. Stephens, C. J. Hewitt, J. R. Powell, and R. A. Badley, "Analysis of bacterial function by multi-colour fluorescence flow cytometry and single cell sorting," Journal of Microbiological Methods, vol. 42, no. 1, pp. 97–114, 2000.
59. J. D. Oliver, "The viable but nonculturable state in bacteria," Journal of Microbiology, vol. 43, pp. 93–100, 2005.
60. B. H. Pyle, S. C. Broadaway, and G. A. McFeters, "Sensitive detection of Escherichia coli O157:H7 in food and water by immunomagnetic separation and solid-phase laser cytometry," Applied and Environmental Microbiology, vol. 65, no. 5, pp. 1966–1972, 1999.
61. M. M. Taimur Khan, B. H. Pyle, and A. K. Camper, "Specific and rapid enumeration of viable but nonculturable and viable-culturable gram-negative bacteria by using flow cytometry," Applied and Environmental Microbiology, vol. 76, no. 15, pp. 5088–5096, 2010.
62. R. Santiso, M. Tamayo, J. Gosálvez, G. Bou, M. C. Fernández, and J. L. Fernández, "A rapid in situ procedure for determination of bacterial susceptibility or resistance to antibiotics that inhibit peptidoglycan biosynthesis," BMC Microbiology, vol. 11, article 191, 2011.
63. M. A. Sánchez-Romero and J. Casadesús, "Contribution of phenotypic heterogeneity to adaptive antibiotic resistance," Proceedings of the

National Academy of Sciences of the United States of America, vol. 111, no. 1, pp. 355–360, 2014.

64. L. Cui, J. Lian, H. Neoh, E. Reyes, and K. Hiramatsu, "DNA microarray-based identification of genes associated with glycopeptide resistance in Staphylococcus aureus," Antimicrobial Agents and Chemotherapy, vol. 49, no. 8, pp. 3404–3413, 2005.
65. H. Xie, J. Mire, Y. Kong et al., "Rapid point-of-care detection of the tuberculosis pathogen using a BlaC-specific fluorogenic probe," Nature Chemistry, vol. 4, no. 10, pp. 802–809, 2012.
66. J. Zhang, Y. Shen, S. L. May, D. C. Nelson, and S. Li, "Ratiometric fluorescence detection of pathogenic bacteria resistant to broad-spectrum β-lactam antibiotics," Angewandte Chemie International Edition, vol. 51, no. 8, pp. 1865–1868, 2012.
67. S. Mizukami, S. Watanabe, Y. Akimoto, and K. Kikuchi, "No-wash protein labeling with designed fluorogenic probes and application to real-time pulse-chase analysis," Journal of the American Chemical Society, vol. 134, no. 3, pp. 1623–1629, 2012.
68. Q. Shao and B. Xing, "Enzyme responsive luminescent ruthenium(ii) cephalosporin probe for intracellular imaging and photoinactivation of antibiotics resistant bacteria," Chemical Communications, vol. 48, no. 12, pp. 1739–1741, 2012.
69. Q. Shao, Y. Zheng, X. Dong, K. Tang, X. Yan, and B. Xing, "A covalent reporter of β-lactamase activity for fluorescent imaging and rapid screening of antibiotic-resistant bacteria," Chemistry, vol. 19, no. 33, pp. 10903–10910, 2013.
70. X.-J. Fu, Y. Fang, and M. Yao, "Antimicrobial photodynamic therapy for methicillin-resistant Staphylococcus aureus infection," BioMed Research International, vol. 2013, Article ID 159157, 9 pages, 2013.
71. H. M. Tang, M. R. Hamblin, and C. M. N. Yow, "A comparative in vitro photoinactivation study of clinical isolates of multidrug-resistant pathogens," Journal of Infection and Chemotherapy, vol. 13, no. 2, pp. 87–91, 2007.
72. P. S. Zolfaghari, S. Packer, M. Singer et al., "In vivo killing of Staphylococcus aureus using a light-activated antimicrobial agent," BMC Microbiology, vol. 9, article 27, 2009.
73. Y. Jiang, A. W. Leung, X. Wang, H. Zhang, and C. Xu, "Inactivation of Staphylococcus aureus by photodynamic action of hypocrellin

B," Photodiagnosis and Photodynamic Therapy, vol. 10, no. 4, pp. 600–606, 2013.

74. A. Aranda, L. Sequedo, L. Tolosa et al., "Dichloro-dihydro-fluorescein diacetate (DCFH-DA) assay: a quantitative method for oxidative stress assessment of nanoparticle-treated cells," Toxicology in Vitro, vol. 27, no. 2, pp. 954–963, 2013.
75. M. Gandhi and M. L. Chikindas, "Listeria: a foodborne pathogen that knows how to survive," International Journal of Food Microbiology, vol. 113, no. 1, pp. 1–15, 2007.
76. M. Utratna, E. Cosgrave, C. Baustian, R. H. Ceredig, and C. P. O'Byrne, "Effects of growth phase and temperature on σB activity within a Listeria monocytogenes population: evidence for RsbV-independent activation of σB at refrigeration temperatures," BioMed Research International, vol. 2014, Article ID 641647, 11 pages, 2014.
77. M. Mols and T. Abee, "Bacillus cereus responses to acid stress," Environmental Microbiology, vol. 13, no. 11, pp. 2835–2843, 2011.
78. U. P. Cronin and M. G. Wilkinson, "Physiological response of Bacillus cereus vegetative cells to simulated food processing treatments," Journal of Food Protection, vol. 71, no. 11, pp. 2168–2176, 2008.
79. M. Mols, R. van Kranenburg, C. C. J. van Melis, R. Moezelaar, and T. Abee, "Analysis of acid-stressed Bacillus cereus reveals a major oxidative response and inactivation-associated radical formation," Environmental Microbiology, vol. 12, no. 4, pp. 873–885, 2010.
80. U. P. Cronin and M. G. Wilkinson, "Monitoring changes in germination and permeability of Bacillus cereus endospores following chemical, heat and enzymatic treatments using flow cytometry," Journal of Rapid Methods and Automation in Microbiology, vol. 16, no. 2, pp. 164–184, 2008.
81. U. P. Cronin and M. G. Wilkinson, "Bacillus cereus endospores exhibit a heterogeneous response to heat treatment and low-temperature storage," Food Microbiology, vol. 25, no. 2, pp. 235–243, 2008.
82. A. Neumeyer, T. Hübschmann, S. Müller, and J. Frunzke, "Monitoring of population dynamics of Corynebacterium glutamicum by multiparameter flow cytometry," Microbial Biotechnology, vol. 6, no. 2, pp. 157–167, 2013.

83. E. O. Sunny-Roberts and D. Knorr, "Evaluation of the response of Lactobacillus rhamnosus VTT E-97800 to sucrose-induced osmotic stress," Food Microbiology, vol. 25, no. 1, pp. 183–189, 2008.
84. A. Reis, T. L. da Silva, C. A. Kent, M. Kosseva, J. C. Roseiro, and C. J. Hewitt, "Monitoring population dynamics of the thermophilic Bacillus licheniformis CCMI 1034 in batch and continuous cultures using multi-parameter flow cytometry," Journal of Biotechnology, vol. 115, no. 2, pp. 199–210, 2005.
85. T. da Silva, A. Reis, C. Kent, M. Kosseva, J. C. Roseiro, and C. Hewitt, "Stress-induced physiological responses to starvation periods as well as glucose and lactose pulses in Bacillus licheniformis CCMI 1034 continuous aerobic fermentation processes as measured by multi-parameter flow cytometry," Biochemical Engineering Journal, vol. 24, no. 1, pp. 11–15, 2005.
86. N. Borth, R. Mitterbauer, D. Mattanovich, W. Kramer, K. Bayer, and H. Katinger, "Flow cytometric analysis of bacterial physiology during induction of foreign protein synthesis in recombinant Escherichia coli cells," Cytometry, vol. 31, no. 2, pp. 125–129, 1998.
87. H. Trip, P. J. van der Veek, T. C. Renniers et al., "A novel screening system for secretion of heterologous proteins in Bacillus subtilis," Microbial Biotechnology, vol. 4, no. 5, pp. 673–682, 2011.
88. H. Sträuber, T. Hübschmann, N. Jehmlich et al., "NBDT (3-(N-(7-nitrobenz-2-oxa-1,3-diazol-4-yl)amino)3-toluene)—a novel fluorescent dye for studying mechanisms of toluene uptake into vital bacteria," Cytometry A, vol. 77, no. 2, pp. 113–120, 2010.
89. R. H. Valdivia, A. E. Hromockyj, D. Monack, L. Ramakrishnan, and S. Falkow, "Applications for green fluorescent protein (GFP) in the study of host-pathogen interactions," Gene, vol. 173, no. 1, pp. 47–52, 1996.
90. S. Dhandayuthapani, L. E. Via, C. A. Thomas, P. M. Horowitz, D. Deretic, and V. Deretic, "Green fluorescent protein as a marker for gene expression and cell biology of mycobacterial interactions with macrophages," Molecular Microbiology, vol. 17, no. 5, pp. 901–912, 1995.
91. L. Kremer, A. Baulard, J. Estaquier, O. Poulain-Godefroy, and C. Locht, "Green fluorescent protein as a new expression marker in mycobacteria," Molecular Microbiology, vol. 17, no. 5, pp. 913–922, 1995.

92. R. H. Valdivia and S. Falkow, "Bacterial genetics by flow cytometry: rapid isolation of Salmonella typhimurium acid-inducible promoters by differential fluorescence induction," Molecular Microbiology, vol. 22, no. 2, pp. 367–378, 1996.
93. S. Aymanns, S. Mauerer, G. van Zandbergen, C. Wolz, and B. Spellerberg, "High-level fluorescence labeling of Gram-positive pathogens," PLoS ONE, vol. 6, no. 6, Article ID e19822, 2011.
94. A. E. Stapleton, C. L. Fennell, D. M. Coder, C. L. Wobbe, P. L. Roberts, and W. E. Stamm, "Precise and rapid assessment of Escherichia coli adherence to vaginal epithelial cells by flow cytometry," Clinical Cytometry, vol. 50, no. 1, pp. 31–37, 2002.
95. B. Hara-Kaonga and T. G. Pistole, "OmpD but not OmpC is involved in adherence of Salmonella entericaserovar Typhimurium to human cells," Canadian Journal of Microbiology, vol. 50, no. 9, pp. 719–727, 2004.
96. B. Hara-Kaonga and T. G. Pistole, "A dual fluorescence flow cytometric analysis of bacterial adherence to mammalian host cells," Journal of Microbiological Methods, vol. 69, no. 1, pp. 37–43, 2007.
97. Z. W. Bent, S. S. Branda, and G. M. Young, "The Yersinia enterocolitica Ysa type III secretion system is expressed during infections both in vitro and in vivo," Microbiologyopen, vol. 2, no. 6, pp. 962–975, 2013.
98. L. P. Barker, D. M. Brooks, and P. L. C. Small, "The identification of Mycobacterium marinum genes differentially expressed in macrophage phagosomes using promoter fusions to green fluorescent protein," Molecular Microbiology, vol. 29, no. 5, pp. 1167–1177, 1998.
99. R. L. Wilson, A. R. Tvinnereim, B. D. Jones, and J. T. Harty, "Identification of Listeria monocytogenes in vivo-induced genes by fluorescence-activated cell sorting," Infection and Immunity, vol. 69, no. 8, pp. 5016–5024, 2001.
100. M. Rathman, N. Jouirhi, A. Allaoui, P. Sansonetti, C. Parsot, and G. T. Van Nhieu, "The development of a FACS-based strategy for the isolation of Shigella flexneri mutants that are deficient in intercellular spread," Molecular Microbiology, vol. 35, no. 5, pp. 974–990, 2000.
101. B. H. Abuaita and J. H. Withey, "Genetic screening for bacterial mutants in liquid growth media by fluorescence-activated cell sorting," Journal of Microbiological Methods, vol. 84, no. 1, pp. 109–113, 2011.

102. J. S. Yu, J. Whitesides, S. H. Lee et al., "Flow cytometry sorting of recombinant mycobacterial species yields bacterial clones with enhanced insert expression," Clinical and Vaccine Immunology, vol. 18, no. 1, pp. 43–49, 2011.

103. E. N. Trip, J.-W. Veening, E. J. Stewart, J. Errington, and D.-J. Scheffers, "Balanced transcription of cell division genes in Bacillus subtilis as revealed by single cell analysis," Environmental Microbiology, vol. 15, no. 12, pp. 3196–3209, 2013.

104. S. Müller and G. Nebe-Von-Caron, "Functional single-cell analyses: Flow cytometry and cell sorting of microbial populations and communities," FEMS Microbiology Reviews, vol. 34, no. 4, pp. 554–587, 2010.

105. S. Müller, "Modes of cytometric bacterial DNA pattern: a tool for pursuing growth," Cell Proliferation, vol. 40, no. 5, pp. 621–639, 2007.

106. C. Jacobs, D. Hung, and L. Shapiro, "Dynamic localization of a cytoplasmic signal transduction response regulator controls morphogenesis during the Caulobacter cell cycle," Proceedings of the National Academy of Sciences of the United States of America, vol. 98, no. 7, pp. 4095–4100, 2001.

107. I. O. Odsbu and K. Skarstad, "Growth rate dependent numbers of SeqA structures organize the multiple replication forks in rapidly growing Escherichia coli," Genes to Cells, vol. 14, no. 5, pp. 643–657, 2009.

108. G. Demarre and D. K. Chattoraj, "DNA adenine methylation is required to replicate both Vibrio choleraechromosomes once per cell cycle.," PLoS genetics, vol. 6, Article ID e1000939, 2010.

109. J. W. Veening, W. K. Smits, L. W. Hamoen, and O. P. Kuipers, "Single cell analysis of gene expression patterns of competence development and initiation of sporulation in Bacillus subtilis grown on chemically defined media," Journal of Applied Microbiology, vol. 101, no. 3, pp. 531–541, 2006.

110. J. D. Chung, G. Stephanopoulos, K. Ireton, and A. D. Grossman, "Gene expression in single cells of Bacillus subtilis: evidence that a threshold mechanism controls the initiation of sporulation," Journal of Bacteriology, vol. 176, no. 7, pp. 1977–1984, 1994.

111. W. K. Smits, C. C. Eschevins, K. A. Susanna, S. Bron, O. P. Kuipers, and L. W. Hamoen, "Stripping Bacillus: comK auto-stimulation is responsible for the bistable response in competence

development," Molecular Microbiology, vol. 56, no. 3, pp. 604–614, 2005.

112. J. Veening, L. W. Hamoen, and O. P. Kuipers, "Phosphatases modulate the bistable sporulation gene expression pattern in Bacillus subtilis," Molecular Microbiology, vol. 56, no. 6, pp. 1481–1494, 2005.
113. W. K. Smits, C. Bongiorni, J. Veening, L. W. Hamoen, O. P. Kuipers, and M. Perego, "Temporal separation of distinct differentiation pathways by a dual specificity Rap-Phr system in Bacillus subtilis," Molecular Microbiology, vol. 65, no. 1, pp. 103–120, 2007.
114. J. W. Veening, O. A. Igoshin, R. T. Eijlander, R. Nijland, L. W. Hamoen, and O. P. Kuipers, "Transient heterogeneity in extracellular protease production by Bacillus subtilis," Molecular Systems Biology, vol. 4, article 184, 2008.
115. U. P. Cronin and M. G. Wilkinson, "The use of flow cytometry to study the germination of Bacillus cereusendospores," Cytometry A, vol. 71, no. 3, pp. 143–153, 2007.
116. J. M. Henke and B. L. Bassler, "Quorum sensing regulates type III secretion in Vibrio harveyi and Vibrio parahaemolyticus," Journal of Bacteriology, vol. 186, no. 12, pp. 3794–3805, 2004.
117. C. Anetzberger, U. Schell, and K. Jung, "Single cell analysis of Vibrio harveyi uncovers functional heterogeneity in response to quorum sensing signals," BMC Microbiology, vol. 12, article 209, 2012.
118. G. O'Toole, H. B. Kaplan, and R. Kolter, "Biofilm formation as microbial development," Annual Review of Microbiology, vol. 54, pp. 49–79, 2000.
119. S. J. Pamp, C. Sternberg, and T. Tolker-Nielsen, "Insight into the microbial multicellular lifestyle via flow-cell technology and confocal microscopy," Cytometry A, vol. 75, no. 2, pp. 90–103, 2009.
120. G. Nebe-Von Caron and R. A. Badley, "Viability assessment of bacteria in mixed populations using flow cytometry," Journal of Microscopy, vol. 179, no. 1, pp. 55–66, 1995.
121. S. S. Branda, Å. Vik, L. Friedman, and R. Kolter, "Biofilms: the matrix revisited," Trends in Microbiology, vol. 13, no. 1, pp. 20–26, 2005.
122. S. S. Branda, J. E. González-Pastor, S. Ben-Yehuda, R. Losick, and R. Kolter, "Fruiting body formation by Bacillus subtilis," Proceedings of the National Academy of Sciences of the United States of America, vol. 98, no. 20, pp. 11621–11626, 2001.

123. U. Bai, I. Mandic-Mulec, and I. Smith, "SinI modulates the activity of SinR, a developmental switch protein of Bacillus subtilis, by protein-protein interaction," Genes and Development, vol. 7, no. 1, pp. 139–148, 1993.

124. D. B. Kearns, F. Chu, S. S. Branda, R. Kolter, and R. Losick, "A master regulator for biofilm formation by Bacillus subtilis," Molecular Microbiology, vol. 55, no. 3, pp. 739–749, 2005.

125. Y. Chai, F. Chu, R. Kolter, and R. Losick, "Bistability and Biofilm formation in Bacillus subtilis," Molecular Microbiology, vol. 67, no. 2, pp. 254–263, 2008.

126. V. L. Marlow, F. R. Cianfanelli, M. Porter, L. S. Cairns, and J. K. Dale, "The prevalence and origin of exoprotease-producing cells in the Bacillus subtilis biofilm," Microbiology, vol. 160, pp. 56–66, 2014.

127. P. B. Beauregard, Y. Chai, H. Vlamakis, R. Losick, and R. Kolter, "Bacillus subtilis biofilm induction by plant polysaccharides," Proceedings of the National Academy of Sciences of the United States of America, vol. 110, no. 17, pp. E1621–E1630, 2013.

128. T. L. da Silva, A. Reis, C. A. Kent, J. C. Roseiro, and C. J. Hewitt, "The use of multi-parameter flow cytometry to study the impact of limiting substrate, agitation intensity, and dilution rate on cell aggregation during Bacillus licheniformis CCMI 1034 aerobic continuous culture fermentations," Biotechnology and Bioengineering, vol. 92, no. 5, pp. 568–578, 2005.

# CHAPTER 9

# IMPEDANCE FLOW CYTOMETRY: A NOVEL TECHNIQUE IN POLLEN ANALYSIS

**Iris Heidmann[1], Grit Schade-Kampmann[2], Joep Lambalk[1] , Marcel Ottiger[2] , Marco Di Berardino[2]**

[1] Enza Zaden, Research and Development B.V. 1600AA Enkhuizen, The Netherlands

[2] Amphasys AG, Technopark Lucerne, 6039 Root D4, Switzerland

## ABSTRACT

### Introduction

An efficient and reliable method to estimate plant cell viability, especially of pollen, is important for plant breeding research and plant production processes. Pollen quality is determined by classical methods, like staining techniques or *in vitro* pollen germination, each having disadvantages with

**Citation:** Heidmann I, Schade-Kampmann G, Lambalk J, Ottiger M, Di Berardino M (2016) Impedance Flow Cytometry: A Novel Technique in Pollen Analysis. PLoS ONE 11(11): e0165531. https://doi.org/10.1371/journal.pone.0165531

respect to reliability, analysis speed, and species dependency. Analysing single cells based on their dielectric properties by impedance flow cytometry (IFC) has developed into a common method for cellular characterisation in microbiology and medicine during the last decade. The aim of this study is to demonstrate the potential of IFC in plant cell analysis with the focus on pollen.

## Method

Developing and mature pollen grains were analysed during their passage through a microfluidic chip to which radio frequencies of 0.5 to 12 MHz were applied. The acquired data provided information about the developmental stage, viability, and germination capacity. The biological relevance of the acquired IFC data was confirmed by classical staining methods, inactivation controls, as well as pollen germination assays.

## Results

Different stages of developing pollen, dead, viable and germinating pollen populations could be detected and quantified by IFC. Pollen viability analysis by classical FDA staining showed a high correlation with IFC data. In parallel, pollen with active germination potential could be discriminated from the dead and the viable but non-germinating population.

## Conclusion

The presented data demonstrate that IFC is an efficient, label-free, reliable and non-destructive technique to analyse pollen quality in a species-independent manner.

# INTRODUCTION

Since the frog-leg experiments of Galvani (1737–1798) and studies in Venus fly-trap by Burdon-Sanderson (1828–1905) numerous invasive and non-invasive techniques to analyse electric signalling across animal and plant tissues/organs have been developed [1,2]. Non-invasive, label-free dielectrophoresis (DEP) techniques take advantage of the reaction of dipolar particles to an applied, non-uniform electric field which allows the characterization of cells based on their conductive and permittive properties [3].

Biological membranes are semi-permeable bilayers of lipids and proteins encapsulating the cellular content of organelles, nucleus and cytoskeleton embedded in a conductive cytoplasm. An intact cellular membrane is not highly conductive and acts electrically as a combination of capacitor and resistor. The capacitance depends on the morphology of the membrane. As higher the morphological complexity, as higher the capacitance and as longer it takes before the membrane is charged. The analysis of different characteristics of single cells depends on the electrical interaction between the cellular surface and the surrounding medium and the permittivity of the cellular content. Therefore, the analysis of surface characteristics like cell size and membrane integrity does not require a high permittivity while the detection of cytoplasmic changes does [4]. Thinner membranes are quickly charged. Damaged membranes leak ions and become conductive which can be detected by specific electrodes. Typically, cell size, shape, and membrane properties are analysed between 10 kHz and 5 MHz, cytoplasmic analyses require frequencies above 10 MHz [4] characterized by the recorded amplitude and phase angle (impedance) at multiple frequencies by an impedance meter [5].

The analysis of electric properties in single cells by impedance flow cytometry (IFC) has been described in microorganisms to estimate viability, membrane potential, as well as cell size [6,7], in mammalian cells to characterize culture conditions and apoptosis [8] and to screen for methods to kill breast cancer cells [9]. The IFC system is based on the Coulter system [10] but uses a microfluidic chip which permits measurements in the radio frequency range from 0.1 to 30 MHz with alternating current (AC). An appropriate parameter to discriminate dead from viable cells proved to be the change of the phase angle of the detected impedance signal, while the amplitude provided information about cell size both displayed in a phase-amplitude dot plot. Depending on the chosen frequency, data of cell size, membrane capacitance, cytoplasmic conductivity of single cells, and cell concentration [7,11] are simultaneously obtained.

Viable and functional pollen grains are essential for sexual plant reproduction, the maintenance of genetic diversity, during breeding processes, and commercial seed production [12]. Developing pollen at a specific microspores stage are the base of numerous *in vitro* methods producing homozygous parental lines [13]. Mature pollen grains that are not directly used for pollination can be conserved by specific methods and stored for off-season crossing or germplasm preservation programs [14].

The characteristic steps of pollen development, pollen germination, and pollen tube formation have been intensively reviewed [15,16]. In brief, the developmental steps are accompanied by changes in cell size, composition of the cytoplasm, cell wall formation, nuclear divisions, and metabolic activity, but seem to occur not necessarily in a highly synchronised manner [17–19]. The carbohydrate type and content is species dependent [20], varies during the development [21], is affected by stress [22,23] and storage conditions [24] which further affects the membrane protective function of sucrose [25] with consequences for pollen viability and later germination capacity.

At dispersal from the anther mature pollen grains are released with a species-specific water content, which classifies pollen as partially hydrated (> 30% water content) or partially dehydrated (< 30% water content) [26].

Rehydration, the first step of the pollen towards germination [27], occurs within 5–60 minutes after contact to the stigma [26]. During this phase the metabolic activity of the pollen grain increases, which is essential for the later pollen tube formation [28]. Pollen tube elongation is controlled by active vesicle and organelle trafficking along an intact cytoskeleton [29]. Interferences, through genetic, environmental or biochemical factors, at any of the developmental steps can lead to aborted, small, shrunken, non-viable or non-germinating pollen [19,27,29,30] with subsequent effects on reproduction [31–35].

The quality of developing and mature pollen is determined by various classical, non-standardized methods, like staining with fluorescent/non-fluorescent dyes [36,37], size measurements [38] or pollen germination. However, each of these methods has its disadvantage with respect to general application, reliability, analysis time, labour intensity, genotype and species dependency in relation to pollen germination [39,40]. Until now, a general method that allows the simultaneous determination of pollen viability and prediction of germination has not been described.

Considering pollen as highly active hydrodynamic units [26,41] that change electric properties of their membranes and cytoplasm as they develop, mature and germinate, IFC should allow a quick and reliable characterisation of these plant cells. The knowledge of the specific hydrodynamic and metabolomic processes of pollen grains and their quality control plays an important role during pollen production, the development of pollen storage protocols, and of cause pollination.

Here we describe for the first time the use of impedance flow cytometry to analyse pollen development and viability in a high-throughput and species independent manner, which also predicts pollen germination.

# MATERIAL AND METHODS

## Sample Preparations

Pollen donor plants (standard lines and breeding material, Enza Zaden, The Netherlands) were grown in heated greenhouses under standard conditions. Developing pollen from microspore (prior pollen mitosis I) to immature stage (after pollen mitosis I) of tobacco (*Nicotiana tabacum*, SR1) or mature pollen (after pollen mitosis II [42]) of cucumber (*Cucumis sativus*) were isolated by gentle squeezing excised anthers directly in IFC measurement buffer (tobacco AF4, cucumber AF5, Amphasys, Switzerland). Dry, mature pollen (pepper (*Capsicum annuum*), tomato (*Solanum lycopersicum*) 'Moneyberg', 'Moneymaker', beef and cherry types) were harvested by shaking open flowers inserted into Eppendorf tubes with an electric tooth brush. The collected pollen were resuspended in IFC buffer (tomato, pepper AF6, Amphasys) and filtrated through 50 μm (tobacco, tomato, pepper) or 150 μm (cucumber) filter units (CellTrics®, Sysmex Partec, Germany) prior analysis. The liquid impedance of the IFC measurement buffers was adjusted to10-30 kΩ.

Viability and developmental stage were measured at 0.5 and/or 12 MHz at the default settings (Trigger level 0.1 V, Modulation 3, Amplifier 6, Demodulation 2, Pump 60 rpm) by loading the pollen samples onto a 120 μm (tobacco, tomato, pepper) or 250 μm (cucumber) channel chip, inserted in an impedance flow cytometer type AmphaZ30 (Amphasys, Switzerland). Data of at least $5 \times 10^3$ cells per sample with a concentration between $5 \times 10^4$ and $10^5$ cells/ml were collected and analysed using AmphaSoft v1.2 and converted to the 2.0 version for a better resolution of the plots (Amphasys, Switzerland).

To determine the optimal conditions to create non-viable controls that allow a clear discrimination between viable and non-viable cells, 100 μl aliquots of pollen suspension in closed 250 μl PCR vessels (Greiner) were exposed to a temperature gradient ranging from 40–60°C for 15 minutes (T100 Thermo cycler, Bio-Rad). Aliquots were cooled down to room temperature and filled up to 1 ml with the same IFC buffer before analysis.).

For cell size determination polystyrene beads of 20, 25 and 30 μm (Sigma) were added in a 1:200 dilution to the filtrated pollen samples.

## Classical Viability and Stage Determination

To compare the IFC data with a classical method pollen viability was determined in parallel to IFC by classical fluorescein diacetate (FDA; Sigma 50 μg/ml in DMSO) staining. For the analysis of the developmental stage microspores from five to six buds were isolated in IFC buffer, precipitated after IFC measurements, stained with 4',6-diamidino-2-phenylindole (DAPI; Sigma [43]), and observed under a fluorescent microscope (Olympus IX70). Images were taken with a DP70 camera using CellSens software (Olympus).

## Detection of Starch Accumulation

Pollen development is accompanied by starch accumulation. For starch detection excised anthers of tobacco flowers were squeezed in Lugol's potassium iodine solution (Sigma [44]) and directly imaged.

## Pollen Germination

Pollen germination was determined on the same sample as used for the IFC analysis. Pollen grains were either recollected after the IFC measurement or the remaining aliquot which was not used for IFC was prepared for the germination assay. Samples were centrifuged at 4000 rpm for five minutes, the IFC buffer was replaced by 200 μl germination buffer (tomato and pepper, Brewbaker and Kwak 1963 [45]; cucumber, Vižintin and Bohanec 2004 [46]), spread into a 125 μl Gene frame (AB-0578, Life Technologies) mounted on a standard glass slide, and incubated for 30 minutes (cucumber) or two (tomato, pepper) hours in closed *in vitro* vent boxes (Duchefa, The Netherlands) lined with wet filter paper (Whatman) at room temperature. The germination data were acquired by analysing at least 200 pollen grains per sample. Inactivation experiments were performed as described above but with a temperature gradient ranging from 20–40°C.

The experiments concerning viability analysis and germination inactivation were performed with three biological replicates collected at different days over a period of two months, the germination prediction was performed with three individual tomato lines collected on the same day.

# RESULTS

## IFC Detects different Developmental Stages

Immature pollen grains (microspores) are essential for the *in vitro* production of homozygous parental lines via the androgenetic pathway. In many doubled haploid protocols the stage between uni-nucleate and early bi-nucleate are the most responsive ones [47]. Their correct stage and concentration of viable cells at the start of the culture is predominantly analysed manually by DAPI and FDA staining, respectively.

To demonstrate that developmental differences can be detected by IFC we isolated microspores from different bud sizes of tobacco and analysed them by nuclear staining (DAPI), potassium iodine (Lugol) staining for starch detection, and IFC at 0.5 and 12 MHz (Fig 1). Along with the increasing tobacco flower size (Fig 1 Buds I—VIII) the progression of pollen development from tetrad to late binucleate stage (Fig 1 DAPI) and starch accumulation (Fig 1Lugol) between stage V and VII was detected. The Amphasoft dot plots of the analysed stages showed two clearly distinguishable populations throughout the tested stages (Fig 1 IFC). While the tetrad stage (Fig 1 Buds I, DAPI) analysed at 0.5 MHz (Fig 1 IFC) show two distinguishable population that vary in size, displayed by the amplitude (y-axis) and phase angle (x-axis), the analysis at 12 MHz shows only one population with strong variation in size and cytoplasmic impedance which is probably a mix of tetrads and release young microspores. Uni-nucleate microspores showed one population at 0.5 MHz but two with a different phase angle when analysed at 12 MHz (Fig 1 Buds II, IFC). All later stages (III–VIII) analysed at 0.5 MHz showed one population with a lower phase angle that remained at the same position throughout the measurements while the second, with the higher phase angle and a shift to the right, definitely changed in both amplitude and phase value (Fig 1 IFC). At the stage III and IV the population with the high phase angle analysed at 12 MHz showed a broad variation from 40° to 90° (Fig 1IFC) which was narrowed at the point of first starch detection at stage V (Fig 1 Lugol and IFC) and subsequent later stages to a phase angle value not higher than 55° (Fig 1 IFC). The comparison of the IFC data with the DAPI staining strongly suggests that the population with the higher and changing phase angle corresponds to developing microspores as it is expected that this group of microspores differs in size and cytoplasmic activity from dead microspores that did not display a change in the phase angle.

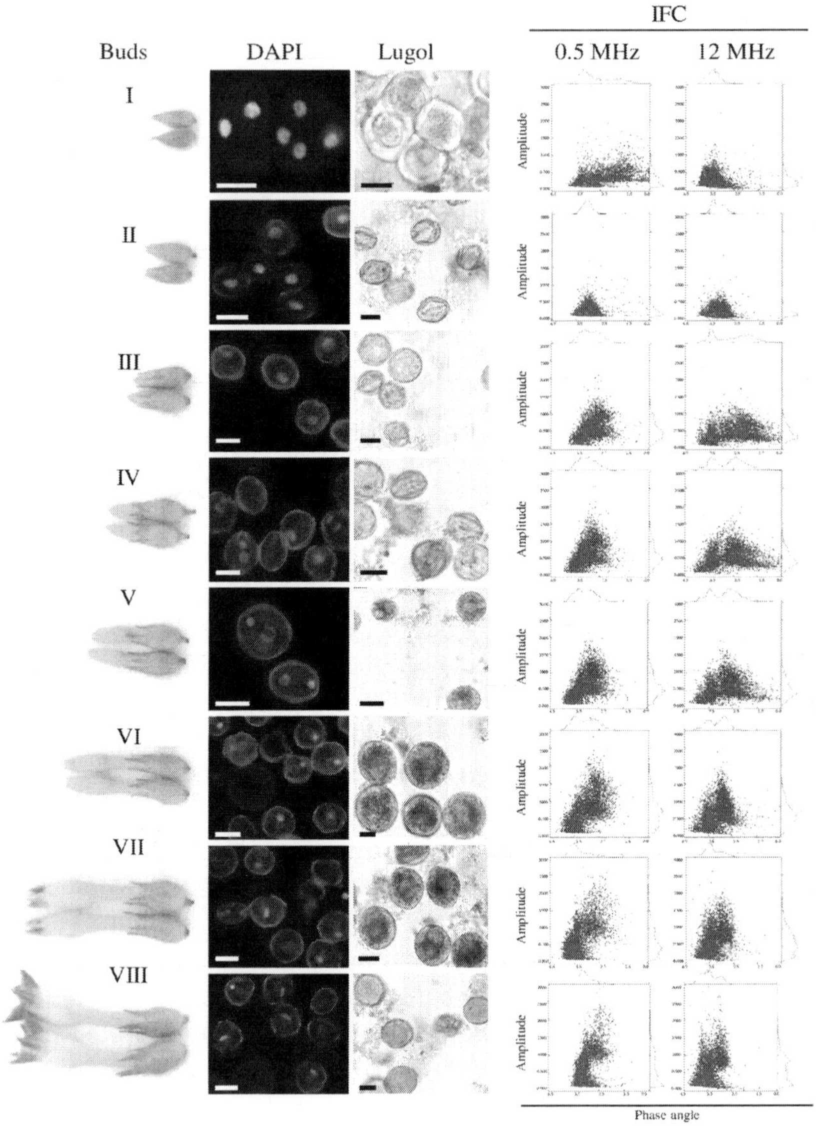


**Figure 1:** Analysis of developing tobacco pollen.

Developmental stages of tobacco pollen according to flower bud size (Buds) was followed by stage determination (DAPI), starch accumulation (Lugol) and IFC analysis at 0.5 and 12 MHz. The stages (I-VIII) represent I, tetrad; II, uni-nucleate, III, late uni-nucleate; VI–VII, binucleate, VIII, mature pollen grains. Scale bars 20 µm

https://doi.org/10.1371/journal.pone.0165531.g001

## IFC Discriminates Efficiently between Viable and Non-viable Pollen Populations

To identify the previously observed populations (Fig 1) we compared freshly harvested, mature pollen of various species with heat-treated aliquots. We found that all untreated pollen samples showed two clearly distinguishable populations with low and high phase angles (Fig 2A, left) while heat-treated pollen showed only one with a low phase angle (Fig 2A, right) which identified the population with the higher phase angle of Fig 2A as the viable one because it is no longer present after heat-inactivation. This pattern was confirmed in all other tested species treated in the same way, in naturally aged pollen e.g. in pollen harvested from old flowers (cucumber four days post anthesis), or pollen (tomato) stored for more than seven days at room temperature. Alternatively, a non-viable pollen population can be determined using polystyrene beads which indicate a certain size but should not show a variation in phase angle.

To demonstrate the suitability of beads, we first analysed beads of 20, 25, 30 µm and a mix thereof with a 120 µm chip at 0.5 and 12 MHz. As expected, the beads could be distinguished according to their size and displayed no change in the phase angle. The bead mix was then added to fresh and heat-treated tomato pollen samples and analysed in the same way. As previously seen in the tobacco samples the fresh sample showed two populations with a different phase angle and variation in size. At 0.5 MHz the left population was almost indistinguishable from the beads, while the population with the high phase angle corresponded to the bead size of 30 µm. At 12 MHz the beads showed a very different phase angle than the two pollen populations. In the heat-treated sample the with the higher phase angle was absent at both, 0.5 and 12 MHz indicating that this particular population must have been the viable one. The intensified signal of the heat-treated pollen population between the signal of 25 and 30 µm beads at 0.5 MHz showed that the heat-treated pollen shifted in both phase angle and size.

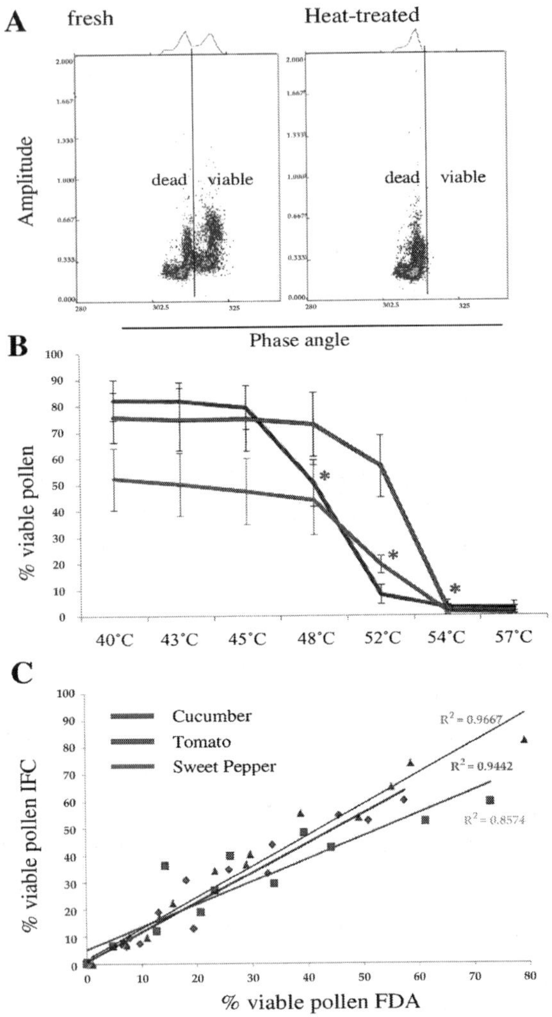


**Figure 2:** IFC discriminates between viable and non-viable pollen populations.

Pollen of three different species, cucumber (green), sweet pepper (blue) and tomato (red) were inactivated by heat. The effect of the heat treatment on pollen viability was analysed by IFC and classical FDA staining. A, AmphaSoft dot plots and histograms at 0.5 MHz of fresh (left) and heat-treated (right) tomato pollen populations; B,Temperature-dependent decrease of pollen viability detected by IFC at 12 MHz, * $p = < 0.05$; C, Correlation of IFC data at 12 MHz with classical FDA staining.

https://doi.org/10.1371/journal.pone.0165531.g002

The gating function of the software allows the automatic calculation of the ratio of dead/viable fractions (line gating) or individual populations (polygon gating) which was not used in the presented images to maintain a clear figure, instead the separation between dead and viable is indicated by a line, the beads are marked in red.

The data demonstrate that dead pollen are not only smaller than viable ones but also behave almost the same way as polystyrene beads in the applied AC field. The observed "J"-shape of the bead and tomato populations at 0.5 MHz is due to the natural size/shape variation of the beads and their individual position within the channel. As the flow rate in the squared microfluidic channel was around 0.5 ml/min (corresponds to a Reynolds number of around 70), the cells are likely aligned along four streams, an effect also known as inertial focussing [48]. In this fluidic situation and within the used electrode configuration the system detects two preferred signal amplitudes, a smaller amplitude for the cells in the middle of the channel, and a larger amplitude for cells on the top or bottom of the channel, which are closer to the electrodes. The slight phase shift leading to the J-shape is an effect of electronic cross-talk between the electrode pairs [49]. This cross-talk is substantial at low frequencies (0.5 MHz), but almost inexistent at high frequencies (12 MHz). During the estimation of the optimal inactivation temperature we found that pollen of different species reacted differently to the applied temperature. The viability of cucumber pollen significantly declined after a 15 minute heat treatment at 48°C while the viability of sweet pepper and tomato pollen declined at higher temperatures. Above 54°C none of the tested pollen was viable anymore (Fig 2B).

We compared our IFC data with classical FDA staining by mixing fresh pollen with heat-treated (60°C) ones in fixed ratio reducing the amount of viable pollen from 100% in steps of 25% down to zero. For the three analysed species, cucumber, sweet pepper, and tomato a high correlation ($R^2$ = 0.85–0.96) between FDA and IFC (Fig 2C) was found.

## IFC Predicts Pollen Germination in Tomato

In parallel to the previously described heat-treatments at 40–60°C we performed pollen germination assays but no pollen germination was observed in this temperature range despite the detected viability (Fig 2B). Therefore, we chose a lower temperature range (20–40°C) to find the critical conditions that would inactivate germination.

When comparing pollen germination of samples before and after measurements we found that pollen germination is not affected by the passage through Ampha Z30. Since the re-collection required more handling we continued the pollen germination assay on the remaining aliquot which was not used for the measurement.

As seen in the earlier experiments, cucumber pollen was more sensitive to a 15 minutes heat treatment than sweet pepper or tomato pollen. Cucumber pollen germination was significantly inhibited at a lower temperature (33°C) than in sweet pepper and tomato (37°C). It was further noticed that with increasing temperatures, above 33°C, pollen tubes shortened and beyond 37°C no pollen germination could be observed (Fig 3A).

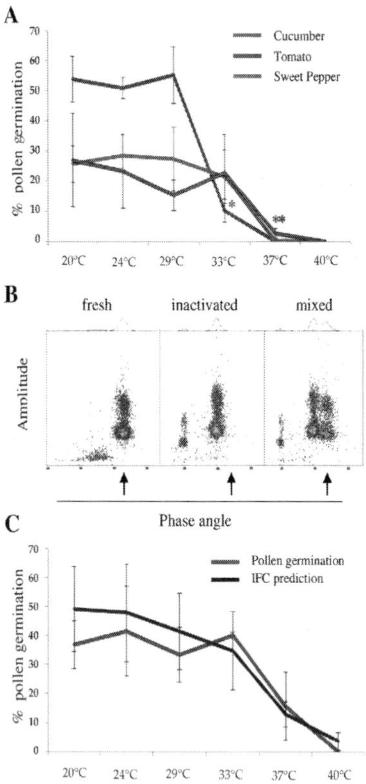


**Figure 3:** Heat inactivation of pollen germination and prediction by IFC.

A, Germination of fresh and heat-inactivated pollen samples, cucumber (green), sweet pepper (blue), and tomato (red), * p = < 0.05; B, AmphaSoft dot plots and histograms of fresh (left), 40°C-inactivated (middle), and a mixed pollen population. The arrow points at the fresh and germinating population C, IFC prediction of tomato pollen germination (purple) and *in vitro* pollen germination (red) after heat-inactivation;

https://doi.org/10.1371/journal.pone.0165531.g003

Tomato pollen can be harvested in a high quantity from a few plants. Therefore, we focussed on tomato to find IFC settings that would predict pollen germination. Pollen samples were collected and treated as described before. To identify the population with active germination capacity we mixed a fresh pollen sample with aliquots of 40°C-inactivated ones of the same genetic background in defined ratios, pre-checked the instrument settings (different MHz) for an optimal resolutions. The previously used frequency of 0.5 MHz was not suitable for the discrimination purposes, so we chose 3 and 12 MHz, the frequencies expected to indicate membrane and cytoplasmic integrity [4], with an adaptation of the settings (Trigger 0.05 V, Modulation 4, Amplifier 6, Demodulation 1). We found that the population with the highest phase angle was missing in the inactivated sample at 3 and 12 MHz and therefore must have been the pollen population with germination capacity. Because of a clearer separation of the inactive and active populations, and minimized undesired electronic cross talk at 12 MHz we used this frequency as standard to predict pollen germination (Fig 3B). We confirmed our findings by exposing freshly harvested pollen of different tomato lines to a temperature gradient (20–40°C) and found no significant difference between the IFC data predicting pollen germination and *in vitro* germination (Fig 3C) when a germination-inactivated control was used as a negative reference.

The high variation per data point is due to the fact that different genotypes were used and that the experiments had to be repeated over several days. It is likely that the greenhouse conditions per day have varied in terms of light intensity and relative humidity, which might have had affected the pollen quality.

## DISCUSSION

IFC has been successfully used for cellular characterisation in bacterial and mammalian systems [6,8,9]. Analysing plant cells and in particular the quality of pollen in an efficient and reliable manner is not only of interest of plant breeding and seed production industries, but plays also a role in scientific projects investigating pollen development [30,50], pollen-related heat or cold tolerance [22,23], male-sterility systems [51], the establishment of pollen storage or doubled haploid protocols [13,14]. The most critical argument against all methods analysing pollen viability is the missing relevance for pollen germination [39]. Viability is a pre-requisite but no warranty for pollen germination, successful fertilisation and fruit set. This cascade of processes depends on external/internal stimuli and specific gene expression of both male and female partner [52–55]. The presented work focusses on viability determination of developing and mature pollen and the prediction of germination.

We have shown that different steps in pollen development can be visualized by IFC. Due to the non-synchronous pollen development and subsequent variations in size and cytoplasmic dynamics [19] the populations are not displayed as a sharp dot but rather as a cloud. The phase angle shift occurring during the binucleate stage (stage VI–VII) is likely caused by an enhanced carbohydrate synthesis and starch accumulation. At maturation stage (VII–VIII) the viable pollen were represented in a compacter cloud than in the stages before suggesting a higher cell synchronisation and a development arrest prior germination [25].

The different populations were identified by heat-treatment or natural aging either at the plant or by wrong storage conditions which caused a decrease in size and shift in phase due to the induced membrane cytoplasmic damages [25]. The discrimination between dead and viable cells can be detected at all frequencies and applied at any stage to all species.

We found that the cells are still functional after the measurement and the passage through the instrument which opens opportunities to expose cells to biotic or abiotic stresses, analyse their properties, and subject them directly to culture, molecular, biochemical characterization, or other purposes like storage or pollination.

The size based stage determination by IFC should also allow the detection of off-type pollen grains as described for the volume-based detection of aneuploid pollen grains [56]. In preliminary experiments we were able to discriminate between diploid and tetraploid tomato individuals of the same genetic background but because of the natural pollen size variation we could not clearly detect a contamination of diploid pollen within a haploid population.

The advantage of the presented method over classical viability methods involving dyes or fluorochromes, is that IFC allows a high throughput viability screening of e.g. large mutant populations, where 5 x $10^3$ cells, depending on the concentration, can be analysed in less than ten seconds in a standardisable manner.

The observation that the technique can discriminate between different stages and can also detect changes in the cytoplasm encouraged us to investigate if pollen germination can be predicted. The dielectric principles of IFC predict an optimal frequency range for cell size analysis between 0.01 and 20 MHz, membrane integrity between 1 and 10 MHz, and cytoplasmic conductivity above 10 MHz [2]. Pollen becomes metabolic active prior germination, thus changes in membrane integrity and cytoplasmic conductivity should be detectable when comparing germinating active with inactive pollen [28].

The finding that we could discriminate between pollen populations with and without germination capacity at 3 and 12 MHz demonstrates that we were able to detect the described metabolic changes pollen germination [19,26–28,41]. Pollen incapable of active re-hydration has a lower water content and likely a slower (passive) up-take than hydrated pollen and subsequently a different resistance when exposed to an electric field, which allows the discrimination between the two populations by IFC and leads to a high predictability of the IFC data for *in vitro*germination as only hydrated pollen is able to germinate [26,27,41]. While the separation of germinating from non-germinating and dead pollen is obvious in the mix of the two differently treated pollen samples (fresh/active versus germination inactivated), the separation is less obvious in a natural population exposed to a temperature gradient. The later can be explained by the expectation that not all pollen have developed synchronously [19]. Subsequently, they will not hydrate, be metabolic activated, and react to heat in a synchronized manner and therefore display a gradient from activated to completely inactivated. Thus, a germination–inactivated control is indispensable for this type of assay.

The effect on heat stress on pollen development and carbohydrate content in tomato was predominantly performed on whole plants [57,58]. Heat treatment applied to hydrated tobacco pollen but not submerged pollen affected the cytoskeleton and sucrose synthase deposition in pollen tubes during germination [29]. We chose to treat the pollen in liquid for an equal distribution of heat during the short-term treatment. The inhibition of tomato pollen germination between 37 and 40°C is in line with previous publications demonstrating that a treatment of 38°C negatively affects pollen germination in tobacco and tomato [29,59,60]. The ability of tomato pollen to germinate after long-term storage or heat stress depends on the biosynthesis of polyamines [60,61], or $\beta$-alanine [62] with a role for specific amino acid transporters during de-/hydration phases [63]. In tobacco (*Nicotiana tabacum*) pollen specific peptides are dephosphorylated during pollen activation [28].

The applied heat stress must have affected one or more described factors responsible for successful pollen germination. While gene expression or enzymatic activities cannot be analysed directly by IFC, consequences thereof can be observed if they affect the cellular integrity and the reaction to an electric field.

The applied combination of controlled and adjustable temperature gradient and IFC analysis opens opportunities to screen for genetic variation in temperature tolerance [22,23].

We have shown that IFC is a quick and reliable method to analyse pollen quality in a species-independent manner. To our knowledge, this is the first demonstration of an easily applicable and non-destructive method that allows the simultaneous discrimination between dead, viable and germinating pollen populations in a single measurement which opens opportunities to screen for any condition that affects pollen development and quality. Future technical IFC developments should allow the separation of the observed populations and in combination with molecular and biochemical methods support the future characterization of specific pollen populations.

## REFERENCES

1. Volkov A (2012) Plant Electrophysiology: Methods and Cell Electrophysiology: Springer Berlin Heidelberg.
2. Cheung KC, Di Berardino M, Schade-Kampmann G, Hebeisen M, Pierzchalski A, et al. (2010) Microfluidic impedance-based flow cytometry. Cytometry Part A 77: 648–666.
3. Yafouz B, Kadri NA, Ibrahim F (2013) Microarray Dot Electrodes Utilizing Dielectrophoresis for Cell Characterization. Sensors 13: 9029–9046. pmid:23857266
4. Sabuncu AC (2011) A microfluidic device for impedance spectroscopy. Norfolk, USA: Old Dominion University.
5. Azzarello E, Masi E, Manusco S (2012) Electrochemical Impedance Spectroscopy. In: Volkov AG, editor. Plant Electrophysiology, Methods and Cell Electrophysiology. Heidelberg, NewYork, Dordrecht, London: Spinger. pp. 205–224.
6. David F, Hebeisen M, Schade G, Franco-Lara E, Di Berardino M (2012) Viability and membrane potential analysis of Bacillus megaterium cells by impedance flow cytometry. Biotechnology and Bioengineering 109: 483–492. pmid:21956238
7. Gawad S, Schild L, Renaud P (2001) Micromachined impedance spectroscopy flow cytometer for cell analysis and particle sizing. Lab on a Chip 1: 76–82. pmid:15100895
8. Pierzchalski A, Hebeisen M, Mittag A, Bocsi J, Di Berardino M, et al. (2012) Label-free hybridoma cell culture quality control by a chip-based impedance flow cytometer. Lab Chip 12: 4533–4543. pmid:22907524
9. Crocetti S, Beyer C, Schade G, Egli M, Fröhlich J, et al. (2013) Low Intensity and Frequency Pulsed Electromagnetic Fields Selectively Impair Breast Cancer Cell Viability. PLoS ONE 8: e72944. pmid:24039828
10. DeBlois RW, Bean CP (1970) Counting and sizing of submicron particles by the resistive pulse technique. Review of Scientific Instruments 41: 909–916.
11. Orczyk W, Przetakiewicz J, Nadolska-Orczyk A (2003) Somatic hybrids of Solanum tuberosum—Application to genetics and breeding. Plant Cell, Tissue and Organ Culture 74: 1–13.

12. Heslop-Harrison JS (1992) Cytological Techniques to Assess Pollen Quality. In: Cresti M, Tiezzi A, editors. Sexual Plant Reproduction: Springer Berlin Heidelberg. pp. 41–48.
13. Maluszynski M, editor (2003) Doubled Haploid Production in Crop Plants: A Manual. Dordrecht/Boston/London: Kluwer.
14. Hanna WW, Towill LE (2010) Long-Term Pollen Storage. In: Janick J, editor. Plant Breeding Reviews. Oxford: John Wiley & Sons. pp. 179–207.
15. Berger F, Twell D (2011) Germline specification and function in plants. Annual Review of Plant Biology. pp. 461–484. pmid:21332359
16. Rutley N, Twell D (2015) A decade of pollen transcriptomics. Plant Reproduction 28: 73–89. pmid:25761645
17. Firon N, Nepi M, Pacini E (2012) Water status and associated processes mark critical stages in pollen development and functioning. Annals of Botany 109: 1201–1213. pmid:22523424
18. Ischebeck T, Valledor L, Lyon D, Gingl S, Nagler M, et al. (2014) Comprehensive cell-specific protein analysis in early and late pollen development from diploid microsporocytes to pollen tube growth. Molecular and Cellular Proteomics 13: 295–310. pmid:24078888
19. Carrizo García C, Nepi M, Pacini E (2016) It is a matter of timing: asynchrony during pollen development and its consequences on pollen performance in angiosperms—a review. Protoplasma: 1–17.
20. Franchi GG, Bellani L, Nepi M, Pacini E (1996) Types of carbohydrate reserves in pollen: Localization, systematic distribution and ecophysiological significance. Flora 191: 143–159.
21. Wei D, Xu H, Li R (2015) Dynamics of polysaccharides and neutral lipids during anther development in castor (Ricinus communis). Journal of the American Society for Horticultural Science 140: 356–361.
22. Müller F, Rieu I (2016) Acclimation to high temperature during pollen development. Plant Reproduction: 1–12.
23. Sharma KD, Nayyar H (2016) Regulatory networks in pollen development under cold stress. Frontiers in Plant Science 7.
24. Geng X, Qiu J, Okubo H (2013) Changes of carbohydrate content during Lilium and Gladiolus pollen cryopreservation. Grana 52: 202–206.
25. Carrizo García C (2007) Pollen starch reserves in tomato relatives: Ecophysiological implications. Grana 46: 13–19.

26. Pacini E, Guarnieri M, Nepi M (2006) Pollen carbohydrates and water content during development, presentation, and dispersal: A short review. Protoplasma 228: 73–77. pmid:16937057
27. Johnson SA, McCormick S (2001) Pollen Germinates Precociously in the Anthers of raring-to-go, an Arabidopsis Gametophytic Mutant. Plant Physiology 126: 685–695. pmid:11402197
28. Fila J, Radau S, Matros A, Hartmann A, Scholz U, et al. (2016) Phosphoproteomics profiling of tobacco mature pollen and pollen activated in vitro. Molecular & Cellular Proteomics 15: 1336–1350.
29. Parrotta L, Faleri C, Cresti M, Cai G (2016) Heat stress affects the cytoskeleton and the delivery of sucrose synthase in tobacco pollen tubes. Planta 243: 43–63. pmid:26335855
30. De Storme N, Geelen D (2014) The impact of environmental stress on male reproductive development in plants: Biological processes and molecular mechanisms. Plant, Cell and Environment 37: 1–18. pmid:23731015
31. Harsant J, Pavlovic L, Chiu G, Sultmanis S, Sage TL (2013) High temperature stress and its effect on pollen development and morphological components of harvest index in the C3 model grass *Brachypodium distachyon*. Journal of Experimental Botany 64: 2971–2983. pmid:23771979
32. Hecker RJ, Stanwood PC, Soulis CA (1986) Storage of sugarbeet pollen. Euphytica 35: 777–783.
33. Herrero MP, Johnson RR (1980) High Temperature Stress and Pollen Viability of Maize. Crop Sci 20: 796–800.
34. Issarakraisila M, Considine JA (1994) Effect of temperature on pollen viability in Mango cv. "Kensington". Annals of Botany 73: 231–240.
35. Zinn KE, Tunc-Ozdemir M, Harper JF (2010) Temperature stress and plant sexual reproduction: Uncovering the weakest links. Journal of Experimental Botany 61: 1959–1968. pmid:20351019
36. Rodriguez-Riano T, Dafni A (2000) A new prodecudre to asses pollen viability. Sexual Plant Reproduction 12: 241–244.
37. Heslop-Harrison J, Heslop-Harrison Y (1970) Evaluation of pollen viability by enzymatically induced fluorescence; intracellular hydrolysis of fluorescein diacetate. Stain technology 45: 115–120. pmid:4192549

38. Kelly JK, Rasch A, Kalisz S (2002) A method to estimate pollen viability from pollen size variation. American Journal of Botany 89: 1021–1023. pmid:21665702
39. Dafni A, Firmage D (2000) Pollen viability and longevity: Practical, ecological and evolutionary implications. Plant Systematics and Evolution 222: 113–132.
40. Johnson-Brousseau SA, McCormick S (2004) A compendium of methods useful for characterizing Arabidopsis pollen mutants and gametophytically-expressed genes. Plant Journal 39: 761–775. pmid:15315637
41. Heslop-Harrison J (1979) An Interpretation of the Hydrodynamics of Pollen. American Journal of Botany 66: 737–743.
42. Twell D (2011) Male gametogenesis and germline specification in flowering plants. Sexual Plant Reproduction 24: 149–160. pmid:21103996
43. Otto F (1990) DAPI Staining of Fixed Cells for High-Resolution Flow Cytometry of Nuclear DNA. Methods in Cell Biology 33: 105–110. pmid:1707478
44. Khan MMR, Hasnunnahar M, Iwayoshi M, Isshiki S (2013) Pollen and seed fertility of the male fertile lines having the fertility restorer gene in three CMS systems of eggplant. Scientia Horticulturae 157: 39–44.
45. Brewbaker JL, Kwack BH (1963) The Essential Role of Calcium Ion in Pollen Germination and Pollen Tube Growth. American Journal of Botany 50: 859–865.
46. Vižintin L, Bohanec B (2004) In vitro manipulation of cucumber (Cucumis sativus L. ) pollen and microspores: Isolation procedures, viability tests, germination, maturation. Acta Biologica Cracoviensia Series Botanica 46: 177–183.
47. Maraschin SF, De Priester W, Spaink HP, Wang M (2005) Androgenic switch: An example of plant embryogenesis from the male gametophyte perspective. Journal of Experimental Botany 56: 1711–1726. pmid:15928015
48. Hansson J, Karlsson JM, Haraldsson T, Wijngaart Wvd, Russom A. Inertial particle focusing in parallel microfluidic channels for high-throughput filtration; 2011 5–9 June 2011. pp. 1777–1780.
49. Spencer D, Caselli F, Bisegna P, Morgan H (2016) High accuracy particle analysis using sheathless microfluidic impedance cytometry.

Lab on a Chip 16: 2467–2473. pmid:27241585

50. Demidov D, Lermontova I, Weiss O, Fuchs J, Rutten T, et al. (2014) Altered expression of Aurora kinases in Arabidopsis results in aneu- and polyploidization. The Plant Journal 80: 449–461. pmid:25146886
51. Chen L, Liu YG (2014) Male sterility and fertility restoration in crops. Annual Review of Plant Biology 65: 579–606. pmid:24313845
52. Steinhorst L, Kudla J (2013) Calcium—a central regulator of pollen germination and tube growth. Biochimica et Biophysica Acta—Molecular Cell Research 1833: 1573–1581.
53. Zechmann B, Koffler BE, Russell SD (2011) Glutathione synthesis is essential for pollen germination in vitro. BMC Plant Biology 11: 54. pmid:21439079
54. Lang V, Pertl-Obermeyer H, Safiarian MJ, Obermeyer G (2014) Pump up the volume—a central role for the plasma membrane H+ pump in pollen germination and tube growth. Protoplasma 251: 477–488. pmid:24097309
55. Matton DP, Nass N, Clarke AE, Newbigin E (1994) Self-incompatibility: how plants avoid illegitimate offspring. Proceedings of the National Academy of Sciences 91: 1992–1997.
56. De Storme N, Zamariola L, Mau M, Sharbel TF, Geelen D (2013) Volume-based pollen size analysis: An advanced method to assess somatic and gametophytic ploidy in flowering plants. Sexual Plant Reproduction 26: 65–81.
57. Firon N, Shaked R, Peet MM, Pharr DM, Zamski E, et al. (2006) Pollen grains of heat tolerant tomato cultivars retain higher carbohydrate concentration under heat stress conditions. Scientia Horticulturae 109: 212–217.
58. Pressman E, Peet M, Pharr DM (2002) The Effect of Heat Stress on Tomato Pollen Characteristics is Associated with Changes in Carbohydrate Concentration in the Developing Anthers. Annals of Botany 90: 631–636. pmid:12466104
59. Shivanna KR, Cresti M (1989) Effects of high humidity and temperature stress on pollen membrane integrity and pollen vigour in Nicotiana tabacum. Sexual Plant Reproduction 2: 137–141.
60. Song J, Nada K, Tachibana S (2002) Suppression of S-adenosylmethionine Decarboxylase Activity is a Major Cause for High-Temperature Inhibition of Pollen Germination and Tube Growth

in Tomato (Lycopersicon esculentum Mill. ). Plant and Cell Physiology 43: 619–627. pmid:12091715

61. Song J, Tachibana S (2007) Loss of viability of tomato pollen during long-term dry storage is associated with reduced capacity for translating polyamine biosynthetic enzyme genes after rehydration. Journal of Experimental Botany 58: 4235–4244. pmid:18065764
62. Shen YH, Chen YH, Liu HY, Chiang FY, Wang YC, et al. (2014) Expression of a gene encoding β-ureidopropionase is critical for pollen germination in tomatoes. Physiologia Plantarum 150: 425–435. pmid:24033314
63. Schwacke R, Grallath S, Breitkreuz KE, Stransky E, Stransky H, et al. (1999) LeProT1, a Transporter for Proline, Glycine Betaine, and γ-Amino Butyric Acid in Tomato Pollen. The Plant Cell 11: 377–391. pmid:10072398

# CHAPTER 10

# FLOW CYTOMETRY BIOINFORMATICS

**Kieran O'Neill[1,2] , Nima Aghaeepour[1,2] , Josef Špidlen[1] , Ryan Brinkman[1,3]**

[1] Terry Fox Laboratory, BC Cancer Agency, Vancouver, British Columbia, Canada

[2] Bioinformatics Graduate Program, University of British Columbia, Vancouver, British Columbia, Canada

[3]Department of Medical Genetics, University of British Columbia, Vancouver, British Columbia, Canada

## ABSTRACT

Flow cytometry bioinformatics is the application of bioinformatics to flow cytometry data, which involves storing, retrieving, organizing, and analyzing flow cytometry data using extensive computational resources and tools. Flow cytometry bioinformatics requires extensive use of and contributes to the development of techniques from computational statistics and machine learning. Flow cytometry and related methods allow the quantification of

**Citation:** O'Neill K, Aghaeepour N, Špidlen J, Brinkman R (2013) Flow Cytometry Bioinformatics. PLoS Comput Biol 9(12): e1003365. https://doi.org/10.1371/journal.pcbi.1003365

multiple independent biomarkers on large numbers of single cells. The rapid growth in the multidimensionality and throughput of flow cytometry data, particularly in the 2000s, has led to the creation of a variety of computational analysis methods, data standards, and public databases for the sharing of results. Computational methods exist to assist in the preprocessing of flow cytometry data, identifying cell populations within it, matching those cell populations across samples, and performing diagnosis and discovery using the results of previous steps. For preprocessing, this includes compensating for spectral overlap, transforming data onto scales conducive to visualization and analysis, assessing data for quality, and normalizing data across samples and experiments. For population identification, tools are available to aid traditional manual identification of populations in two-dimensional scatter plots (gating), to use dimensionality reduction to aid gating, and to find populations automatically in higher dimensional space in a variety of ways. It is also possible to characterize data in more comprehensive ways, such as the density-guided binary space partitioning technique known as probability binning, or by combinatorial gating. Finally, diagnosis using flow cytometry data can be aided by supervised learning techniques, and discovery of new cell types of biological importance by high-throughput statistical methods, as part of pipelines incorporating all of the aforementioned methods.

Open standards, data, and software are also key parts of flow cytometry bioinformatics. Data standards include the widely adopted Flow Cytometry Standard (FCS) defining how data from cytometers should be stored, but also several new standards under development by the International Society for Advancement of Cytometry (ISAC) to aid in storing more detailed information about experimental design and analytical steps. Open data is slowly growing with the opening of the CytoBank database in 2010 and FlowRepository in 2012, both of which allow users to freely distribute their data, and the latter of which has been recommended as the preferred repository for MIFlowCyt-compliant data by ISAC. Open software is most widely available in the form of a suite of Bioconductor packages, but is also available for web execution on the GenePattern platform.

## FLOW CYTOMETRY DATA

Flow cytometers operate by hydrodynamically focusing suspended cells so that they separate from each other within a fluid stream. The stream is passed by one or more lasers, and the resulting fluorescent and scattered light is detected by photomultipliers. By using optical filters, particular fluorophores on or within the cells can be quantified by peaks in their emission spectra. This process is illustrated in Figure 1. Reporter molecules may be endogenous fluorophores such as chlorophyll or transgenic green fluorescent protein, or they may be artificial fluorophores covalently bonded to detection molecules such as antibodies for detecting proteins, or hybridization probes for detecting DNA or RNA.

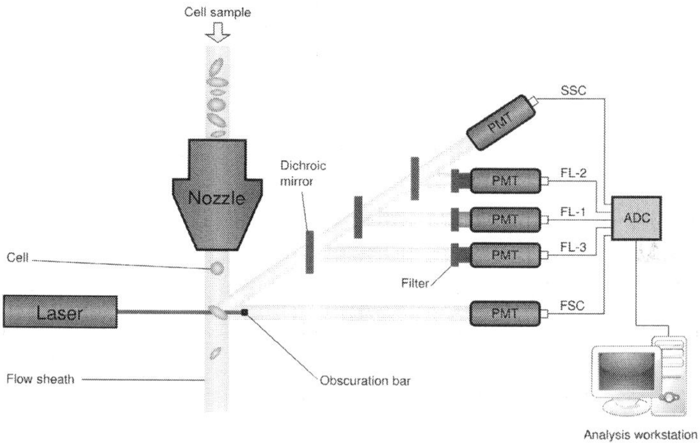


**Figure 1:** Schematic diagram of a flow cytometer, showing focusing of the fluid sheath, laser, optics (in simplified form, omitting focusing), photomultiplier tubes (PMTs), analogue-to-digital converter, and analysis workstation.

https://doi.org/10.1371/journal.pcbi.1003365.g001

The ability to quantify these has led to flow cytometry being used in a wide range of applications, including but not limited to:

- Monitoring of CD4 count in HIV [1]
- Diagnosis of various cancers [2], [3]
- Analysis of aquatic microbiomes [4]
- Sperm sorting [5]
- Measuring telomere length [6]

Until the early 2000s, flow cytometry could only measure a few fluorescent markers at a time. Through the late 1990s into the mid-2000s, however, rapid development of new fluorophores resulted in modern instruments capable of quantifying up to 18 markers per cell [7]. More recently, the new technology of mass cytometry replaces fluorophores with rare earth elementsdetected by time of flight mass spectrometry, achieving the ability to measure the expression of 34 or more markers [8]. At the same time, microfluidic qPCR methods are providing a flow cytometry–like method of quantifying 48 or more RNA molecules per cell [9]. The rapid increase in the dimensionality of flow cytometry data coupled with the development of high-throughput robotic platforms capable of assaying hundreds to thousands of samples automatically have created a need for improved computational analysis methods [7].

## STEPS IN COMPUTATIONAL FLOW CYTOMETRY DATA ANALYSIS

The process of moving from primary FCM data to disease diagnosis and biomarker discovery (illustrated in Figure 2) involves four major steps:

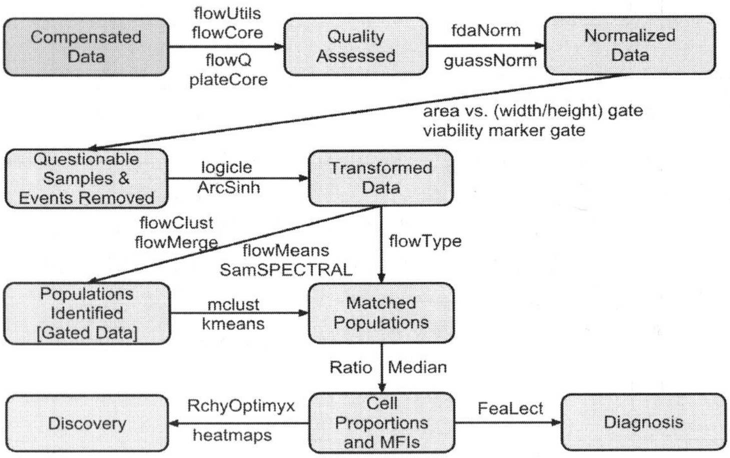


**Figure 2:** An example pipeline for analysis of FCM data and some of the Bioconductor packages relevant to each step.

https://doi.org/10.1371/journal.pcbi.1003365.g002

- Data preprocessing (including compensation, transformation, and normalization)

- Cell population identification (a.k.a. gating)
- Cell population matching for cross-sample comparison
- Relating cell populations to external variables (diagnosis and discovery)

Saving of the steps taken in a particular flow cytometry workflow is supported by some flow cytometry software, and is important for the reproducibility of flow cytometry experiments. However, saved workspace files are rarely interchangeable between software [10]. An attempt to solve this problem is the development of the Gating-ML XML-based data standard (discussed in more detail in the data formats and interchange section), which is slowly being adopted in both commercial and open-source flow cytometry software [11].

# DATA PREPROCESSING

## Compensation

When more than one fluorochrome is used with the same laser, their emission spectrafrequently overlap. Each particular fluorochrome is typically measured using a bandpass optical filter set to a narrow band at or near the fluorochrome's emission intensity peak. The result is that the reading for any given fluorochrome is actually the sum of that fluorochrome's peak emission intensity, and the intensity of all other fluorochromes' spectra where they overlap with that frequency band. This overlap is termed spillover, and the process of removing spillover from flow cytometry data is called compensation [12].

Compensation is typically accomplished by running a series of representative samples each stained for only one fluorochrome, to give measurements of the contribution of each fluorochrome to each channel [12]. The total signal to remove from each channel can be computed by solving a system of linear equations based on this data to produce a spillover matrix, which when inverted and multiplied with the raw data from the cytometer produces the compensated data [12], [13]. The processes of computing the spillover matrix, or applying a precomputed spillover matrix to compensate flow cytometry data, are standard features of flow cytometry software [14].

## Transformation

Cell populations detected by flow cytometry are often described as having approximately log-normal expression [15]. As such, they have traditionally been transformed to a logarithmic scale. In early cytometers, this was often accomplished even before data acquisition by use of a log amplifier. On modern instruments, data is usually stored in linear form, and transformed digitally prior to analysis.

However, compensated flow cytometry data frequently contains negative values due to compensation, and cell populations do occur that have low means and normal distributions [16]. Logarithmic transformations cannot properly handle negative values, and poorly display normally distributed cell types [16], [17]. Alternative transformations that address this issue include the log-linear hybrid transformations Logicle [16], [18] and Hyperlog [19], as well as the hyperbolic arcsine and the Box-Cox [20].

A comparison of commonly used transformations concluded that the biexponential and Box-Cox transformations, when optimally parameterized, provided the clearest visualization and least variance of cell populations across samples [17]. However, a later comparison of the flowTrans package used in that comparison indicated that it did not parameterize the Logicle transformation in a manner consistent with other implementations, potentially calling those results into question [21].

## Quality Control

Particularly in newer, high-throughput experiments, there is a need for visualization methods to help detect technical errors in individual samples. One approach is to visualize summary statistics, such as the empirical distribution functions of single dimensions of technical or biological replicates to ensure they are the similar [22]. For more rigor, the KolmogorovSmirnov test can be used to determine if individual samples deviate from the norm [22]. The Grubbs test for outliers may be used to detect samples deviating from the group.

A method for quality control in higher-dimensional space is to use probability binning with bins fit to the whole dataset pooled together [23]. Then the standard deviation of the number of cells falling in the bins within each sample can be taken as a measure of multidimensional similarity, with samples that are closer to the norm having a smaller standard deviation [23]. With this method, higher standard deviation can indicate outliers, although this is a relative measure as the absolute value depends partly on the number of bins.

With all of these methods, the cross-sample variation is being measured. However, this is the combination of technical variations introduced by the instruments and handling, and actual biological information that is desired to be measured. Disambiguating the technical and the biological contributions to between-sample variation can be a difficult to impossible task [24].

### Normalization

Particularly in multicenter studies, technical variation can make biologically equivalent populations of cells difficult to match across samples. Normalization methods to remove technical variance, frequently derived from image registration techniques, are thus a critical step in many flow cytometry analyses. Single-marker normalization can be performed using landmark registration, in which peaks in a kernel density estimate of each sample are identified and aligned across samples [24].

## IDENTIFYING CELL POPULATIONS

A critical step in analysis of flow cytometric data is the identification of multidimensional regions that contain functionally and phenotypically homogeneous groups of cells for further analysis [27].

### Gating

The data generated by flow cytometers can be plotted in one or two dimensions to produce a histogram or scatter plot. The regions on these plots can be sequentially separated, based on fluorescence intensity, by creating a series of subset extractions, termed "gates." These gates can be produced using software, e.g., FlowJo [28], FCS Express [29], WinMDI [30], CytoPaint (aka Paint-A-Gate) [31], VenturiOne, CellQuest Pro, Cytospec [32], or Kaluza [33].

In datasets with a low number of dimensions and limited cross-sample technical and biological variability (e.g., clinical laboratories), manual analysis of specific cell populations can produce effective and reproducible results. However, exploratory analysis of a large number of cell populations in a high-dimensional dataset is not feasible [34]. In addition, manual analysis in less controlled settings (e.g., cross-laboratory studies) can increase the overall error rate of the study [35]. In one study, several computational gating algorithms performed better than manual analysis in the presence of some variation [26] (illustrated in Figure 3). However, despite the considerable advances in computational analysis, manual gating remains the main solution for the identification of specific rare cell populations that are not well-separated from other cell types.

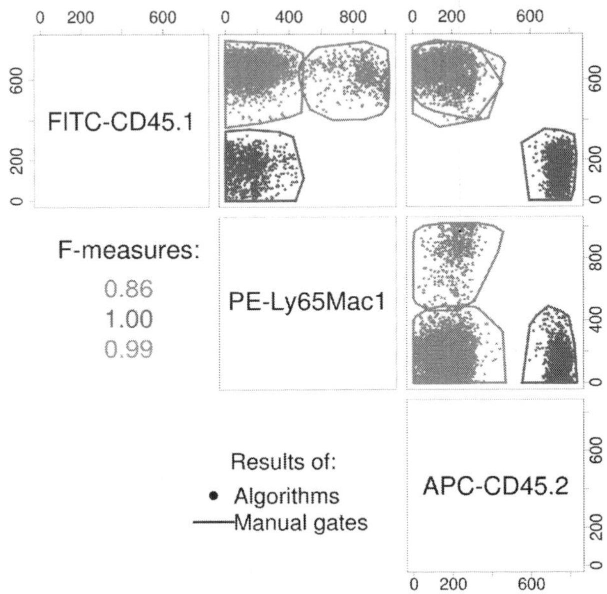


**Figure 3:** Comparison of consensus of eight independent manual gates (polygons) and automated gates (colored dots).

The consensus of the manual gates and the algorithms were produced using the CLUE package [25]. Figure reproduced with permission from [26]. https://doi.org/10.1371/journal.pcbi.1003365.g003

## *Gating Guided by Dimension Reduction*

As the number of markers measured by flow cytometry increases, the number of scatter plots that need to be investigated increases exponentially (some markers need to be investigated several times for each group of cells to resolve high-dimensional differences between cell types that appear to be similar in most markers) [36]. To address this issue, principal component analysis has been used to summarize the high-dimensional datasets using a combination of markers that maximizes the variance of all data points [37]. However, PCA is a linear method and is not able to preserve complex and non-linear relationships. More recently, two-dimensional minimum spanning tree layouts have been used to guide the manual gating process (illustrated in Figure 4). Density-based down-sampling and clustering was used to better represent rare populations and control the time and memory complexity of the minimum spanning tree construction process [38]. More sophisticated dimension reduction algorithms are yet to be investigated [39].

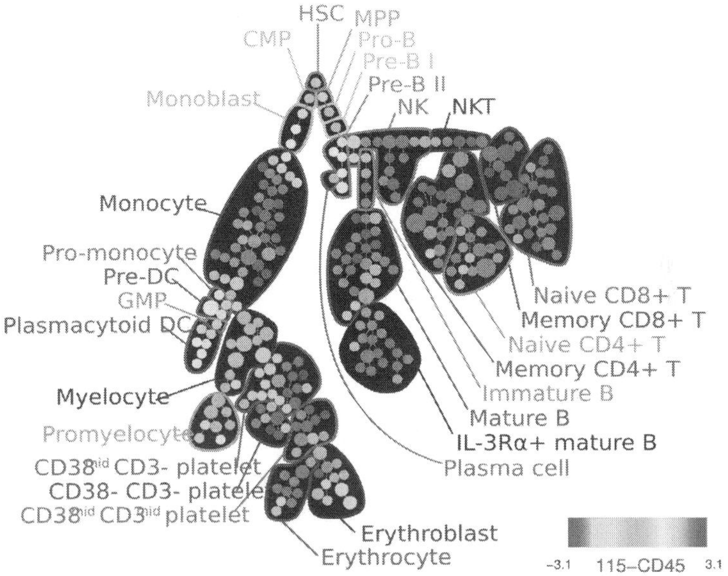


**Figure 4:** Cell populations in a high-dimensional mass-cytometry dataset manually gated after dimension reduction using 2-D layout for a minimum spanning tree.

Figure reproduced from the data provided in [40].

https://doi.org/10.1371/journal.pcbi.1003365.g004

## *Automated Gating*

Developing computational tools for identification of cell populations has been an area of active research only since 2008. Many individual clustering approaches have recently been developed, including model-based algorithms (e.g., flowClust [41] and FLAME [42]), density-based algorithms (e.g., FLOCK [43] and SWIFT), graph-based approaches (e.g., SamSPECTRAL

[44]), and, most recently, hybrids of several approaches (flowMeans [45] and flowPeaks [46]). These algorithms are different in terms of memory and time complexity, their software requirements, their ability to automatically determine the required number of cell populations, and their sensitivity and specificity. The FlowCAP (Flow Cytometry: Critical Assessment of Population Identification Methods) project, with active participation from most academic groups with research efforts in the area, is providing a way to objectively cross-compare state-of-the-art automated analysis approaches [26].

## Probability Binning Methods

Probability binning is a non-gating analysis method in which flow cytometry data is split into quantiles on a univariate basis (shown in Figure 5) [47]. The locations of the quantiles can then be used to test for differences between samples using the chi-squared test [47].

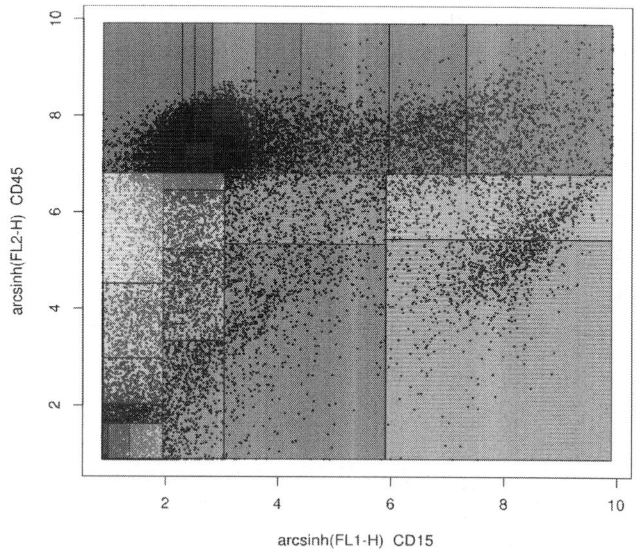


**Figure 5:** An example of probability binning, created using the flowFP Bioconductor package.

The dots represent individual events in an FCS file. The rectangles represent the bins.

https://doi.org/10.1371/journal.pcbi.1003365.g005

This was later extended into multiple dimensions in the form of frequency difference gating, a binary space partitioning technique in which data is iteratively partitioned along the median [48]. These partitions (or bins) are fit to a control sample. Then the proportion of cells falling within each bin in test samples can be compared to the control sample by the chi-squared test.

Finally, cytometric fingerprinting uses a variant of frequency difference gating to set bins and measure for a series of samples how many cells fall within each bin [23]. These bins can be used as gates and used for subsequent analysis similarly to automated gating methods.

### Combinatorial Gating

High-dimensional clustering algorithms are often unable to identify rare cell types that are not well separated from other major populations. Matching these small cell populations across multiple samples is even more challenging. In manual analysis, prior biological knowledge (e.g., biological controls) provides guidance to reasonably identify these populations. However, integrating this information into the exploratory clustering process (e.g., as in semi-supervised learning) has not been successful.

An alternative to high-dimensional clustering is to identify cell populations using one marker at a time and then combine them to produce higher-dimensional clusters. This functionality was first implemented in FlowJo [28]. The flowType algorithm builds on this framework by allowing the exclusion of the markers [49]. This enables the development of statistical tools (e.g., RchyOptimyx) that can investigate the importance of each marker and exclude high-dimensional redundancies [50].

## DIAGNOSIS AND DISCOVERY

After identification of the cell population of interest, a cross-sample analysis can be performed to identify phenotypical or functional variations that are correlated with an external variable (e.g., a clinical outcome). These studies can be partitioned into two main groups:

## Diagnosis

In these studies, the goal usually is to diagnose a disease (or a sub-class of a disease) using variations in one or more cell populations. For example, one can use multidimensional clustering to identify a set of clusters, match them across all samples, and then use supervised learning to construct a classifier for prediction of the classes of interest (e.g., this approach can be used to improve the accuracy of the classification of specific lymphoma subtypes [51]). Alternatively, all the cells from the entire cohort can be pooled into a single multidimensional space for clustering before classification [52]. This approach is particularly suitable for datasets with a high amount of biological variation (in which cross-sample matching is challenging) but requires technical variations to be carefully controlled [53].

## Discovery

In a discovery setting, the goal is to identify and describe cell populations correlated with an external variable (as opposed to the diagnosis setting in which the goal is to combine the predictive power of multiple cell types to maximize the accuracy of the results). Similar to the diagnosis use case, cluster matching in high-dimensional space can be used for exploratory analysis, but the descriptive power of this approach is very limited, as it is hard to characterize and visualize a cell population in a high-dimensional space without first reducing the dimensionality [52], [54]. Finally, combinatorial gating approaches have been particularly successful in exploratory analysis of FCM data. Simplified Presentation of Incredibly Complex Evaluations (SPICE) is a software package that can use the gating functionality of FlowJo to statistically evaluate a wide range of different cell populations and visualize those that are correlated with the external outcome. flowType and RchyOptimyx (as previously discussed) expand this technique by adding the ability of exploring the impact of independent markers on the overall correlation with the external outcome. This enables the removal of unnecessary markers and provides a simple visualization of all identified cell types. In a recent analysis of a large (n=466) cohort of HIV+ patients, this pipeline identified three correlates of protection against HIV, only one of which had been previously identified through extensive manual analysis of the same dataset (as illustrated in Figure 6) [49].

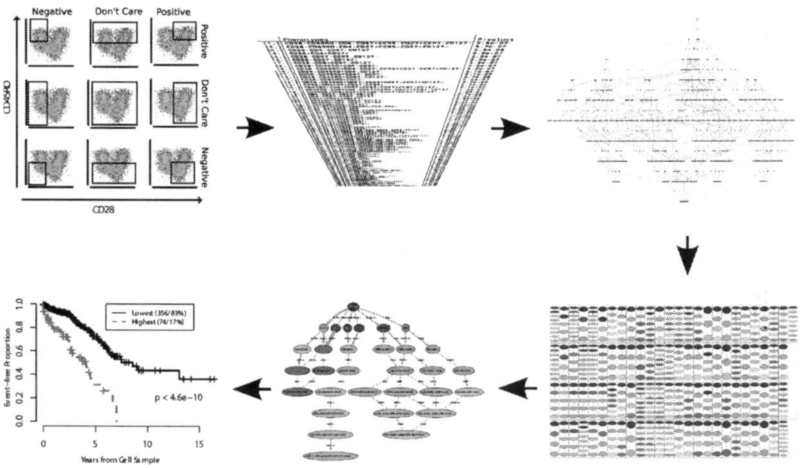


**Figure 6:** Overview of the flowType/RchyOptimyx pipeline for identification of correlates of protection against HIV.

First, tens of thousands of cell populations are identified by combining one-dimensional partitions (panel 1). The cell populations are then analyzed using a statistical test (and Bonferroni's method for multiple testing correction) to identify those correlated with the survival information. Panel 3 shows a complete gating hierarchy describing all possible strategies for gating that cell population. This graph can be mined to identify the "best" gating strategy (i.e., the one in which the most important markers appear earlier). These hierarchies for all selected phenotypes are demonstrated in panel 4. In panel 5, these hierarchies are merged into a single graph that summarizes the entire dataset and demonstrates the trade-off between the number of markers involved in each phenotype and the significance of the correlation with the clinical outcome (e.g., as measured by the KaplanMeier estimator in panel 6). Figure reproduced in part from [49] (public domain) and [50].

https://doi.org/10.1371/journal.pcbi.1003365.g006

# DATA FORMATS AND INTERCHANGE

## Flow Cytometry Standard

Flow cytometry data is typically saved for analysis in the form of an array, with fluorescence and scatter channels represented in columns and individual "events" (most of which are cells) forming the rows, as shown in Figure 7. The number of events acquired from each sample usually ranges between the low thousands and the low millions.

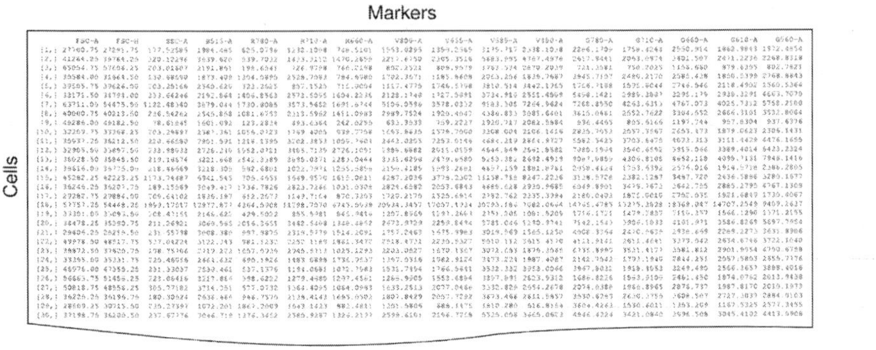


**Figure 7:** Representation of flow cytometry data from an instrument with three scatter channels and 13 fluorescent channels.

Only the values for the first 30 (of hundreds of thousands) of cells are shown.

https://doi.org/10.1371/journal.pcbi.1003365.g007

The first version of a Flow Cytometry Standard (FCS) was developed in 1984 [55]. Since then, FCS has become the standard file format supported by all flow cytometry software and hardware vendors. FCS is a binary file format with three main segments: a text segment containing metadata in keyword/value pairs structures, a data segment usually containing a matrix of detected expression values (so-called list mode format), and a rarely used analysis segment. The FCS specification has traditionally been developed and maintained by the International Society for Advancement of Cytometry (ISAC) [56].

Over the years, updates have been incorporated to adapt to technological advancements in both flow cytometry and computing technologies.

In 1990, FCS 2.0 was introduced [57]. Features introduced in FCS 2.0 included the option of multiple datasets within a data file, the use of different byte orders accommodating hardware variations on different computing platforms, and basic compensation and scaling information. FCS 2.0 was followed by FCS 3.0 in 1997, which introduced the possibility of storing datasets larger than 100 MB [58].

The latest version, FCS 3.1, was introduced in 2010 [59]. It retains the basic FCS file structure and most features of previous versions of the standard. Changes included in FCS 3.1 address potential ambiguities in the previous versions and provide a more robust standard. They include simplified support for international characters and improved support for storing compensation. The major additions are support for preferred display scale, a standardized way of capturing the sample volume, information about the origins of the data file, and support for plate and well identification in high-throughput, plate-based experiments.

FCS used to be the only widely adopted file format in flow cytometry. Recently, additional standard file formats have been developed by ISAC.

## netCDF

ISAC is considering replacing FCS with a flow cytometry–specific version of the Network Common Data Form (netCDF) file format [60]. netCDF is a set of freely available software libraries and machine-independent data formats that support the creation, access, and sharing of array-oriented scientific data. In 2008, ISAC drafted the first version of netCDF conventions for storage of raw flow cytometry data [61].

## Archival Cytometry Standard (ACS)

The Archival Cytometry Standard (ACS) is being developed to bundle data with different components describing cytometry experiments [62]. It captures relations among data, metadata, analysis files, and other components, and includes support for audit trails, versioning, and digital signatures. The ACS container is based on the ZIP file format with an XML-based table of contents specifying relations among files in the container. The XML Signature W3C Recommendation has been adopted to allow for digital signatures of components within the ACS container. An initial draft of ACS was designed in 2007 and finalized in 2010. Since then, ACS support has been introduced in several software tools including FlowJo and Cytobank.

## Gating-ML

The lack of gating interoperability has traditionally been a bottleneck preventing reproducibility of flow cytometry data analysis and the usage of multiple analytical tools. To address this shortcoming, ISAC developed Gating-ML, an XML-based mechanism to formally describe gates and related data (scale) transformations [10]. The draft recommendation version of Gating-ML was approved by ISAC in 2008, and it is partially supported by tools like FlowJo, the flowUtils library in R/Bioconductor, and FlowRepository [62]. It supports rectangular gates, polygon gates, convex polytopes, ellipsoids, decision trees, and Boolean collections of any of the other types of gates. In addition, it includes dozens of built-in public transformations that have been shown to be potentially useful for display or analysis of cytometry data. In 2013, Gating-ML version 2.0 was approved by ISAC›s Data Standards Task Force as a Recommendation. This new version offers slightly less flexibility in terms of the power of gating description; however, it is also significantly easier to implement in software tools [11].

### Classification Results (CLR)

The Classification Results (CLR) File Format [63] has been developed to exchange the results of manual gating and algorithmic classification approaches in a standard way in order to be able to report and process the classification. CLR is based on the commonly supported CSV file format with columns corresponding to different classes and cell values containing the probability of an event being a member of a particular class. These are captured as values between 0 and 1. Simplicity of the format and its compatibility with common spreadsheet tools have been the major requirements driving the design of the specification. Although it was originally designed for the field of flow cytometry, it is applicable in any domain that needs to capture either fuzzy or unambiguous classifications of virtually any kinds of objects.

## PUBLIC DATA AND SOFTWARE

As in other bioinformatics fields, development of new methods has primarily taken the form of free open-source software, and several databases have been created for depositing open data.

### Bioconductor

The Bioconductor project is a repository of free open-source software, mostly written in the R programming language [64]. As of July 2013, Bioconductor contained 21 software packages for processing flow cytometry data [65]. These packages cover most of the range of functionality described earlier in this article.

## GenePattern

GenePattern is a predominantly genomic analysis platform with over 200 tools for analysis of gene expression, proteomics, and other data. A web-based interface provides easy access to these tools and allows the creation of automated analysis pipelines enabling reproducible research. Recently, a GenePattern Flow Cytometry Suite has been developed in order to bring advanced flow cytometry data analysis tools to experimentalists without programmatic skills. It contains close to 40 open-source GenePattern flow cytometry modules covering methods from basic processing of flow cytometry standard (i.e., FCS) files to advanced algorithms for automated identification of cell populations, normalization, and quality assessment. Internally, most of these modules leverage from functionality developed in Bioconductor.

Much of the functionality of the Bioconductor packages for flow cytometry analysis has been packaged up for use with the GenePattern [66] workflow system, in the form of the GenePattern Flow Cytometry Suite [67].

## Public Databases

The Minimum Information about a Flow Cytometry Experiment (MIFlowCyt) requires that any flow cytometry data used in a publication be available, although this does not include a requirement that it be deposited in a public database [68]. Thus, although the journals Cytometry A and B, as well as all journals from the Nature Publishing Group require MIFlowCyt compliance, there is still relatively little publicly available flow cytometry data. Some efforts have been made toward creating public databases, however.

Firstly, CytoBank, which is a complete web-based flow cytometry data storage and analysis platform, has been made available to the public in a limited form [69]. Using the CytoBank code base, FlowRepository was developed in 2012 with the support of ISAC to be a public repository of flow cytometry data [70]. FlowRepository facilitates MIFlowCyt compliance [71], and as of July 2013 contained 65 public datasets [72].

## Datasets

In 2012, the flow cytometry community started to release a set of publicly available datasets. A subset of these datasets representing the existing data analysis challenges is described below. For comparison against manual gating, the FlowCAP-I project has released five datasets, manually gated by human analysts, and two of them gated by eight independent analysts [26]. The FlowCAP-II project included three datasets for binary classification and also reported several algorithms that were able to classify these samples perfectly. FlowCAP-III included two larger datasets for comparison against manual gates as well as one more challenging sample classification dataset. As of March 2013, public release of FlowCAP-III was still in progress [73]. The datasets used in FlowCAP-I, II, and III either have a low number of subjects or parameters. However, recently several more complex clinical datasets have been released including a dataset of 466 HIV-infected subjects, which provides both 14 parameter assays and sufficient clinical information for survival analysis [50], [74]–[76].

Another class of datasets are higher-dimensional mass cytometry assays. A representative of this class of datasets is a study that includes analysis of two bone marrow samples using more than 30 surface or intracellular markers under a wide range of different stimulations [8]. The raw data for this dataset is publicly available as described in the manuscript, and manual analyses of the surface markers are available upon request from the authors.

## OPEN PROBLEMS

Despite rapid development in the field of flow cytometry bioinformatics, several problems remain to be addressed.

Variability across flow cytometry experiments arises from biological variation among samples, technical variations across instruments used, as well as methods of analysis. In 2010, a group of researchers from Stanford University and the National Institutes of Health pointed out that while technical variation can be ameliorated by standardizing sample handling, instrument setup, and choice of reagents, solving variation in analysis methods will require similar standardization and computational automation of gating methods [77]. They further opined that centralization of both data and analysis could aid in decreasing variability between experiments and in comparing results [77].

This was echoed by another group of Pacific Biosciences and Stanford University researchers, who suggested that cloud computing could enable centralized, standardized, high-throughput analysis of flow cytometry experiments [78]. They also emphasized that ongoing development and adoption of standard data formats could continue to aid in reducing variability across experiments [78]. They also proposed that new methods will be needed to model and summarize results of high-throughput analysis in ways that can be interpreted by biologists [78], as well as ways of integrating large-scale flow cytometry data with other high-throughput biological information, such as gene expression, genetic variation, metabolite levels, and disease states [78].

## ACKNOWLEDGMENTS

The version history of the text file and the peer reviews (and response to reviews) are available as supporting information in Text S1 and Text S2.

## REFERENCES


1. Brando B, Barnett D, Janossy G, Mandy F, Autran B, et al. (2000) Cytofluorometric methods for assessing absolute numbers of cell subsets in blood. European Working Group on Clinical Cell Analysis. Cytometry 42: 327–346.
2. Ferreira-Facio CS, Milito C, Botafogo V, Fontana M, Thiago LS, et al. (2013) Contribution of multiparameter flow cytometry immunophenotyping to the diagnostic screening and classification of pediatric cancer. PLoS ONE 8: e55534
3. Wu D, Wood BL, Fromm JR (2013) Flow cytometry for non-Hodgkin and classical Hodgkin lymphoma. Methods Mol Biol 971: 27–47.
4. Wang Y, Hammes F, De Roy K, Verstraete W, Boon N (2010) Past, present and future applications of flow cytometry in aquatic microbiology. Trends Biotechnol 28: 416–424.
5. Johnson LA, Flook JP, Look MV, Pinkel D (1987) Flow sorting of X and Y chromosome-bearing spermatozoa into two populations. Gamete Res 16: 1–9.
6. Baerlocher GM, Vulto I, de Jong G, Lansdorp PM (2006) Flow cytometry and FISH to measure the average length of telomeres (flow FISH). Nat Protoc 1: 2365–2376.
7. Chattopadhyay PK, Hogerkorp C-M, Roederer M (2008) A chromatic explosion: the development and future of multiparameter flow cytometry. Immunology 125: 441–449.
8. Behbehani GK, Bendall SC, Clutter MR, Fantl WJ, Nolan GP (2012) Single-cell mass cytometry adapted to measurements of the cell cycle. Cytometry A 81: 552–566.
9. White AK, VanInsberghe M, Petriv OI, Hamidi M, Sikorski D, et al. (2011) High-throughput microfluidic single-cell RT-qPCR. Proc Natl Acad Sci U S A 108: 13999–14004.
10. Spidlen J, Leif RC, Moore W, Roederer M (2008) International Society for the Advancement of Cytometry Data Standards Task Force, (2008) et al. Gating-ML: XML-based gating descriptions in flow cytometry. Cytometry A 73A: 1151–1157.
11. International Society for Advancement of Cytometry (2013) Gating-ML 2. 0 (Report). Available: http://flowcyt. sourceforge. net/gating/latest. pdf. Accessed 8 November 2013.

12. Roederer M (2002) Compensation in flow cytometry. Curr Protoc Cytom Chapter 1: Unit 1. 14.
13. Bagwell CB, Adams EG (1993) Fluorescence spectral overlap compensation for any number of flow cytometry parameters. Ann N Y Acad Sci 677: 167–184.
14. Hahne F, LeMeur N, Brinkman RR, Ellis B, Haaland P, et al. (2009) flowCore: a Bioconductor package for high throughput flow cytometry. BMC Bioinformatics 10: 106.
15. Shapiro HM (2003) Practical flow cytometry. New York: Wiley-Liss. pp. 235.
16. Parks DR, Roederer M, Moore WA (2006) A new "Logicle" display method avoids deceptive effects of logarithmic scaling for low signals and compensated data. Cytometry A 69: 541–551.
17. Finak G, Perez J-M, Weng A, Gottardo R (2010) Optimizing transformations for automated, high throughput analysis of flow cytometry data. BMC Bioinformatics 11: 546.
18. Moore WA, Parks DR (2012) Update for the logicle data scale including operational code implementations. Cytometry A 81: 273–277.
19. Bagwell CB (2005) Hyperlog-a flexible log-like transform for negative, zero, and positive valued data. Cytometry A 64: 34–42.
20. Lo K, Brinkman RR, Gottardo R (2008) Automated gating of flow cytometry data via robust model-based clustering. Cytometry A 73: 321–332.
21. Qian Y, Liu Y, Campbell J, Thomson E, Kong YM, et al. (2012) FCSTrans: an open source software system for FCS file conversion and data transformation. Cytometry A 81: 353–356.
22. Le Meur N, Rossini A, Gasparetto M, Smith C, Brinkman RR, et al. (2007) Data quality assessment of ungated flow cytometry data in high throughput experiments. Cytometry A 71: 393–403.
23. Rogers WT, Moser AR, Holyst HA, Bantly A, Mohler ER, et al. (2008) Cytometric fingerprinting: quantitative characterization of multivariate distributions. Cytometry A 73: 430–441.
24. Hahne F, Khodabakhshi AH, Bashashati A, Wong C-J, Gascoyne RD, et al. (2010) Per-channel basis normalization methods for flow cytometry data. Cytometry A 77: 121–131.
25. Hornik K, Böhm W (2013) clue: Cluster ensembles, version 0. 3-47. Available: http://cran. r-project. org/web/packages/clue/index. html.

Accessed 15 February 2013.

26. Aghaeepour N, Finak G (2013) FlowCAP Consortium DREAM Consortium (2013) Hoos H, et al. (2013) Critical assessment of automated flow cytometry data analysis techniques. Nat Methods 10: 228–238.
27. Lugli E, Roederer M, Cossarizza A (2010) Data analysis in flow cytometry: the future just started. Cytometry A 77: 705–713.
28. Tree Star, Inc (2013) FlowJo. Available: http://www. flowjo. com/vX/en/ws. boolean. html. Accessed 5 April 2013.
29. De Novo Software (2013) FCS Express. Available: http://www. denovosoftware. com/site/Flow-RUO-Overview. shtml. Accessed 3 April 2013.
30. Saluk A (2013) TSRI Cytometry Software Page. Available: http://facs. scripps. edu/software. html. Accessed September 3 2009.
31. Leukobyte (2013) CytoPaint Classic. Available: http://leukobyte. com Accessed 5 April 2013.
32. Purdue University Cytometry Laboratories (2013) PUCL Cytometry Software Page. Available: http://www. cyto. purdue. edu/Purdue_software. Accessed 7 July 2011.
33. Beckman Coulter (2013) Kaluza. Available: http://www. beckmancoulter. com. Accessed 10 February 2013.
34. Bendall SC, Nolan GP (2012) From single cells to deep phenotypes in cancer. Nat Biotechnol 30: 639–647.
35. Maecker HT, Rinfret A, D'Souza P, Darden J, Roig E, et al. (2005) Standardization of cytokine flow cytometry assays. BMC Immunol 6: 13.
36. Virgo PF, Gibbs GJ (2012) Flow cytometry in clinical pathology. Ann Clin Biochem 49: 17–28.
37. Costa ES, Pedreira CE, Barrena S, Lecrevisse Q, Flores J, et al. (2010) Automated pattern-guided principal component analysis vs expert-based immunophenotypic classification of B-cell chronic lymphoproliferative disorders: a step forward in the standardization of clinical immunophenotyping. Leukemia 24: 1927–1933.
38. Qiu P, Simonds EF, Bendall SC, Gibbs KD, Bruggner RV, et al. (2011) Extracting a cellular hierarchy from high-dimensional cytometry data with SPADE. Nat Biotechnol 29: 886–891.

39. van der Maaten L (2013) Matlab Toolbox for Dimensionality Reduction. Available: http://homepage. tudelft. nl/19j49/Matlab_Toolbox_for_Dimensionality_Reduction. html. Accessed 10 February 2013.

40. Bendall SC, Simonds EF, Qiu P, Amir E-aD, Krutzik PO, et al. (2011) Single-cell mass cytometry of differential immune and drug responses across a human hematopoietic continuum. Science 332: 687–696.

41. Lo K, Hahne F, Brinkman RR, Gottardo R (2009) flowClust: a Bioconductor package for automated gating of flow cytometry data. BMC Bioinformatics 10: 145.

42. Pyne S, Hu X, Wang K, Rossin E, Lin T-I, et al. (2009) Automated high-dimensional flow cytometric data analysis. Proc Natl Acad Sci U S A 106: 8519–8524.

43. Qian Y, Wei C, Eun-Hyung Lee F, Campbell J, Halliley J, et al. (2010) Elucidation of seventeen human peripheral blood B-cell subsets and quantification of the tetanus response using a density-based method for the automated identification of cell populations in multidimensional flow cytometry data. Cytometry B Clin Cytom 78 Suppl 1: S69–82.

44. Zare H, Shooshtari P, Gupta A, Brinkman RR (2010) Data reduction for spectral clustering to analyze high throughput flow cytometry data. BMC Bioinformatics 11: 403.

45. Aghaeepour N, Nikolic R, Hoos HH, Brinkman RR (2011) Rapid cell population identification in flow cytometry data. Cytometry A 79: 6–13.

46. Ge Y, Sealfon SC (2012) flowPeaks: a fast unsupervised clustering for flow cytometry data via K-means and density peak finding. Bioinformatics 28: 2052–2058.

47. Roederer M, Treister A, Moore W, Herzenberg LA (2001) Probability binning comparison: a metric for quantitating univariate distribution differences. Cytometry 45: 37–46.

48. Roederer M, Hardy RR (2001) Frequency difference gating: a multivariate method for identifying subsets that differ between samples. Cytometry 45: 56–64.

49. Aghaeepour N, Chattopadhyay PK, Ganesan A, O'Neill K, Zare H, et al. (2012) Early immunologic correlates of HIV protection can be identified from computational analysis of complex multivariate T-cell flow cytometry assays. Bioinformatics 28: 1009–1016.

50. Aghaeepour N, Jalali A, O'Neill K, Chattopadhyay PK, Roederer M,

et al. (2012) RchyOptimyx: cellular hierarchy optimization for flow cytometry. Cytometry A 81: 1022–1030.

51. Zare H, Bashashati A, Kridel R, Aghaeepour N, Haffari G, et al. (2012) Automated analysis of multidimensional flow cytometry data improves diagnostic accuracy between mantle cell lymphoma and small lymphocytic lymphoma. Am J Clin Pathol 137: 75–85.
52. Qiu P (2012) Inferring phenotypic properties from single-cell characteristics. PLoS ONE 7: e37038
53. Bodenmiller B, Zunder ER, Finck R, Chen TJ, Savig ES, et al. (2012) Multiplexed mass cytometry profiling of cellular states perturbed by small-molecule regulators. Nat Biotechnol 30: 858–867.
54. Bashashati A, Johnson NA, Khodabakhshi AH, Whiteside MD, Zare H, et al. (2012) B cells with high side scatter parameter by flow cytometry correlate with inferior survival in diffuse large B-cell lymphoma. Am J Clin Pathol 137: 805–814.
55. Murphy RF, Chused TM (1984) A proposal for a flow cytometric data file standard. Cytometry 5: 553–555.
56. International Society for Advancement of Cytometry (2013) International Society for Advancement of Cytometry. Available: http:// isac-net. org/. Accessed 5 March 2013.
57. Dean PN, Bagwell CB, Lindmo T, Murphy RF, Salzman GC (1990) Introduction to flow cytometry data file standard. Cytometry 11: 321–322.
58. Seamer LC, Bagwell CB, Barden L, Redelman D, Salzman GC, et al. (1997) Proposed new data file standard for flow cytometry, version FCS 3. 0. Cytometry 28: 118–122.
59. Spidlen J, Moore W, Parks D, Goldberg M, Bray C, et al. (2010) Data File Standard for Flow Cytometry, version FCS 3. 1. Cytometry A 77: 97–100.
60. Leif RC, Spidlen J, Brinkman RR (2009) Cytometry Standards Continuum [7264-16]. In: SPIE BIOS Proc. Vol. 6859. 8–12 February 2009. Available: http://newportinstrument. com/cytometryml/pdf/ Cytometry_Stds_Continuum08. pdf. Accessed 5 March 2013.
61. International Society for Advancement of Cytometry (2008) Analytical Cytometry Standard NetCDF Conventions for List Mode Binary Data File Component. Available: http://flowcyt. sourceforge. net/netcdf/ latest. pdf. Accessed 5 March 2013.

62. Spidlen J, Shooshtari P, Kollmann TR, Brinkman RR (2011) Flow cytometry data standards. BMC Res Notes 4: 50.

63. International Society for Advancement of Cytometry (2012) Classification Results File Format (Report). Available: http://flowcyt. sourceforge. net/clr/latest. pdf. Accessed 5 March 2013.

64. Gentleman RC, Carey VJ, Bates DM, Bolstad B, Dettling M, et al. (2004) Bioconductor: open software development for computational biology and bioinformatics. Genome Biol 5: R80.

65. Bioconductor (2013) Bioconductor FlowCytometry view. Available: http://www. bioconductor. org/packages/release/BiocViews. html#___ FlowCytometry. Accessed 11 July 2013.

66. Reich M, Liefeld T, Gould J, Lerner J, Tamayo P, et al. (2006) GenePattern 2. 0. Nat Genet 38: 500–501.

67. Reich M, Liefeld T, Gould J, Lerner J, Tamayo P, et al. . (2012) GenePattern Flow Cytometry Suite. Available: http://www. broadinstitute. org/cancer/software/genepattern/modules/flow_ cytometry. Accessed 11 July 2013.

68. Lee JA, Spidlen J, Boyce K, Cai J, Crosbie N, et al. (2008) MIFlowCyt: the minimum information about a Flow Cytometry Experiment. Cytometry A 73: 926–930.

69. Kotecha N, Krutzik PO, Irish JM (2010) Web-based analysis and publication of flow cytometry experiments. Curr Protoc Cytom Chapter 10: Unit 10. 17.

70. Spidlen J, Breuer K, Rosenberg C, Kotecha N, Brinkman RR (2012) FlowRepository: a resource of annotated flow cytometry datasets associated with peer-reviewed publications. Cytometry A 81: 727–731.

71. Spidlen J, Breuer K, Brinkman R (2012) Preparing a Minimum Information about a Flow Cytometry Experiment (MIFlowCyt) compliant manuscript using the International Society for Advancement of Cytometry (ISAC) FCS file repository (FlowRepository. org). Curr Protoc Cytom Chapter 10: Unit 10. 18.

72. Spidlen J (2013) Listing of FlowRepository public data sets. Available: http://flowrepository. org/public_experiment_representations. Accessed 5 July 2013.

73. FlowCAP - Flow Cytometry: Critical Assessment of Population Identification Methods. http://flowcap. flowsite. org. Accessed 15 March 2013.

74. Aghaeepour N (2012) IDCRP's HIV Natural History Study data set. Available: https://flowrepository. org/id/FR-FCM-ZZZK. Accessed 3 March 2013.

75. Craig FE, Brinkman RR, Eyck ST, Aghaeepour N (2013) Computational analysis optimizes the flow cytometric evaluation for lymphoma. Cytometry B Clin Cytom E-pub ahead of print. doi:https://doi. org/10. 1002/cyto. b. 21115.

76. Villanova F, Di Meglio P, Inokuma M, Aghaeepour N, Perucha E, et al. (2013) Integration of lyoplate based flow cytometry and computational analysis for standardized immunological biomarker discovery. PLoS ONE 8: e65485

77. Maecker HT, McCoy JP, Nussenblatt R (2012) Standardizing immunophenotyping for the Human Immunology Project. Nat Rev Immunol 12: 191–200.

78. Schadt EE, Linderman MD, Sorenson J, Lee L, Nolan GP (2010) Computational solutions to large-scale data management and analysis. Nat Rev Genet 11: 647–657.

# INDEX

## D



## E



## F